LONG DISTANCE
TELEVISION RECEPTION (TV-DX)
FOR THE
ENTHUSIAST

LONG DISTANCE
TELEVISION RECEPTION (TV-DX)
FOR THE
ENTHUSIAST

by

ROGER W. BUNNEY

BERNARD BABANI (publishing) LTD
THE GRAMPIANS
SHEPHERDS BUSH ROAD
LONDON W6 7NF
ENGLAND

© 1978, © 1979 BERNARD BABANI (publishing) LTD

First Published October 1978

Revised and Reprinted January 1981

British Library Cataloguing in Publication Data
Bunney, Roger W
    Long distance television reception (TV-DX) for
    the enthusiast. — (Bernards & Babani Press
    radio and electronics books; 52).
    1. Television — Receivers and reception — Amateurs'
    manuals
    I. Title
    621. 388'36   TK9960

    ISBN 0 900162 71 6

Printed and bound in Great Britain by
Cox & Wyman Ltd, Reading

PREFACE

Unlike its counterpart Short Wave radio, the art of long distance television reception had by its very specialised and technical nature remained an activity with a very small following until recent years. In the early 1960's the magazine 'Practical Television' first published a regular column intended for exponents of this hobby, edited by a true enthusiast, Charles Rafarel. Under Charles' inspired writing the hobby expanded into a considerable interest and in 1971 plans were made for a descriptive pamphlet covering all aspects of long distance television reception. Unfortunately Charles died before the project was completed.

IPC Magazines invited this author to continue the monthly column and over the past ten years the following has increased still further and continues to attract more enthusiasts each month. Long distance television reception (TV-DX) is a hobby that requires a degree of technical knowledge and competence, a hobby that can still allow a degree of original research and experimentation.

Between 1972 and 1976 there were published three editions of the 'Long Distance Television' booklet, presenting the basic theory and practical aspects of the hobby to both experienced and would-be enthusiasts. There has been considerable technical development since the 3rd. edition of the 'Long Distance Television' booklet and quite dramatic advances are likely to occur in the next decade particularly with reception techniques. I am grateful therefore to Bernard Babani (publishing) Ltd for publishing an enlarged and updated volume presenting both established and new operational methods and practice.

This volume first appeared in late 1978 and in the first 18 months of its life changes within the broadcasting field merited an updated and expanded 2nd. edition. Developments in the direct to home broadcasting satellite suggest that by the mid 1980s a service of sorts will be established for parts of Western Europe, most likely West Germany in the first instance. Consequently the satellite section has been enlarged to encompass further generalised information since in the coming decade direct to home transmissions will perhaps herald the most dramatic changes in the broadcasting field, both from a technical and political viewpoint. In May 1980 the UK broadcasting authorities announced a scheduled time-table for the closure of the 405 line VHF network, commencing with little used low powered relays in 1982 and terminating with the main high powered transmitters in early 1986. The future of the Band 1 spectrum within the UK is uncertain after this date but possibly may be exploited for mobile communications, assuming that the broadcasters have no wish to re-engineer VHF for some form of radio/TV transmissions. WARC '79 allocated further spectrum at the high frequency ends of both Band 3 and Band 5 to provide additional TV channels and the Band 2 VHF sound radio channels will be increased and eventually taking over the complete 88-108MHz spectrum, now in part occupied by various utilities for communication purposes. A further development in terrestrial broadcasting is the use of the 12GHz band for 'fill-in' relays into very small pockets of difficult reception. This technique is currently being exploited by Japan particularly in built up areas of dense population where a single large building may shadow a very localised area, the transmitter often being sited atop the obstruction itself! Stereo sound is also being used on a regular (daily) basis for TV programming by both NHK and commercial broadcasters in Japan.

All the units and devices described within these pages have been designed and used by active enthusiasts, often considerable ingenuity and thought has gone into the development of such units in surmounting individual problems.

Finally, my thanks are due to the many friends that have co-operated and assisted with the production of this volume — without whose help much of the information would not have been available.

Roger Bunney — January 1981

CONTENTS

Page

ACKNOWLEDGEMENTS

The author wishes to thank the following organisations and individuals for their assistance and co-operation in the preparation of this volume.

Independent Broadcasting Authority, Winchester.
'Television', IPC Magazines Ltd., London (particularly for their permission in using certain material published in 'Television' over the past 12 years).

Aerialite Ltd., Oldham, Lancs.
Antiference Ltd. Aylesbury, Bucks.
Hugh Cocks Television Services, Robertsbridge, E. Sussex.
Jaybeam Aerials Ltd., Northampton.
Premier Industries (Cheltenham) Ltd., Cheltenham, Glos.
South West Aerial Systems, Shaftesbury, Dorset.
Wolsey Electronics., Porth, Glamorgan.

Clive Athowe, Norwich.
Steven Birkill, Sheffield.
Cliff Dykes, Orpington.
David Bunyan, Sittingbourne.
Graham Deaves, Cambridge.
Keith Hammer, Derby.
Peter Jansen, Amsterdam, Holland.
Anthony Mann, Perth, Australia.
Ryn Muntjewerff, Beemster, Holland.
Alan Latham, Carshalton.
Garry Smith, Derby.
John Tellick, Surbiton.
Andrew Tett, Tolworth.
David Martin, Shaftesbury.
Christopher Wilson, Potters Bar.
Allan Latham, Abu Dhabi.

INTRODUCTION

As most television viewers will be aware, only a limited
number of stations are available to them at reasonable
strengths in their particular area. Viewers in the United
Kingdom normally have available 3 programme channels (4
channels as of November 1982) transmitted within a UHF channel
grouping from a common transmitting site. Such a transmitting
site will be a high powered 'main' transmitter usually located
upon elevated ground, or a low powered relay serving a restricted
area that is shielded from the main transmitter and hence requiring an improved signal field strength. Examples of both a
main and relay transmitters are shown in the accompanying photographs. Tuning onto other channels will often reaveal somewhat
weaker signals from adjoining transmission areas. At some
favourable high locations a number of alternative programmes
may be available at fair signal strengths, thus enthusing the
viewer to erect suitable aerials to enable reception of stronger
and more consistent signals.

Television signals are transmitted at VHF (Very High
Frequency) or UHF (Ultra High Frequency). Such frequencies
must be used for television due to the extremely wide bandwidths required to transmit high definition vision and sound
information. Due to the use of VHF and UHF, television
transmissions are limited to the area surrounding the transmitter,
extending to the optical horizon and a little beyond at lower
field strengths. The depth of the latter area will depend upon a
number of factors, the main one being the frequency involved,
e.g. a greater area for a transmitter operating at 45MHz, than
say a transmitter operating at 800MHz. To provide as large a
coverage area as possible from the transmitter, a high transmitting mast is used, with high gain aerials often in use to favour
one direction in preference to another direction or area. The
field strength delivered to a receiving site will therefore depend
upon a number of factors: the frequency of the transmitter's
effective radiating power (erp), the height of transmitting
mast and the intervening terrain between the transmitter and
receiving site. Thus at some distance beyond the optical horizon

Emley Moor transmitting site
– an IBA/BBC 'main' station with erp of 870 kw

it would at first seem that distant signals would be too weak to
be of any use in providing viewable pictures. However it is
certainly possible to obtain reception of such transmissions at
very considerable distances, often at high signal strengths.

*Cop Hill transmitting mast
– an IBA relay transmitter of 1 kw erp*

With the advent of the communications satellite and direct
satellite-to-home TV transmissions new techniques in signal
reception and demodulation are being evolved. The successful
ATS-6 experiment with educational broadcasts to the Indian

sub-continent show that it is possible to receive and display
TV pictures in areas some thousands of miles from the target
area and on modified domestic apparatus. Initial experiments
have been carried out at 860MHz., 4GHz., and 12GHz., and
shortly will see TV transmissions from synchronous
orbiting satellites operating within the 11.7–12.5GHz bands.

The mechanism by which these signals are propagated and
received will be discussed later but initially it is necessary to
detail the various television transmission systems at present in
use throughout the world.

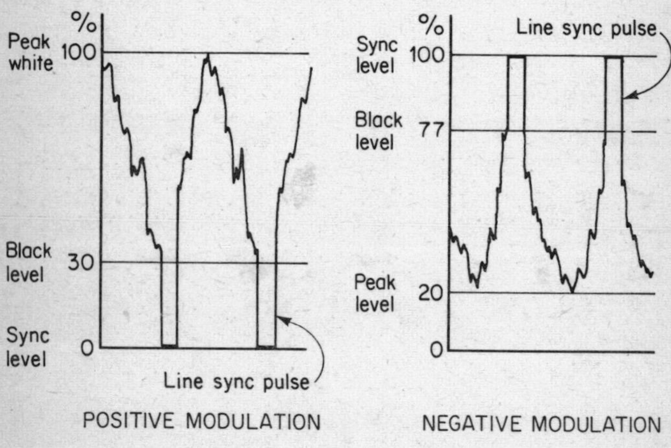

Video waveform showing positive and negative modulation

World Transmission Standards

System	Line No.	Overall channel bandwidth (MHz)	Vision bandwidth (MHz)	Sound/vision spacing (MHz)	Vision modulation	Sound modulation	Areas in use
A	405	5	3	−3.5	+	AM	UK (VHF)
B	625	7	5	+5.5	−	FM	Western Europe, parts of Africa, Middle East, Australasia (VHF)
C	625	7	5	+5.5	+	AM	Luxembourg (VHF)
D	625	8	6	+6.5	−	FM	Eastern Europe, USSR, China (VHF/UHF)
E	819	14	10	±11.15	+	AM	France, Monaco (VHF). Possible future change to System L on VHF.
F	819	7	5	+5.5	+	AM	
G/H	625	8	5	+5.5	−	FM	Western Europe (UHF). System H has a 1.25 MHz vestigal sideband used in Belgium (UHF)
I	625	8	5.5	+6.0	−	FM	UK (UHF), Eire (VHF/UHF), Republic of South Africa (VHF/UHF).
K	625	8	6	+6.5	−	FM	French Territories overseas.
L	625	8	6	+6.5	+	AM	France, Luxembourg (UHF).
M	525	6	4.2	+4.5	−	FM	North & South America, Caribbean, parts of Pacific, Far East, US Forces broad—casting (AFRTS), Japan.

N.B.   It is anticipated that the United Kingdom System A and France System E transmissions will terminate by 1986

Channel Allocations European Bands 1 and 2 (TV)

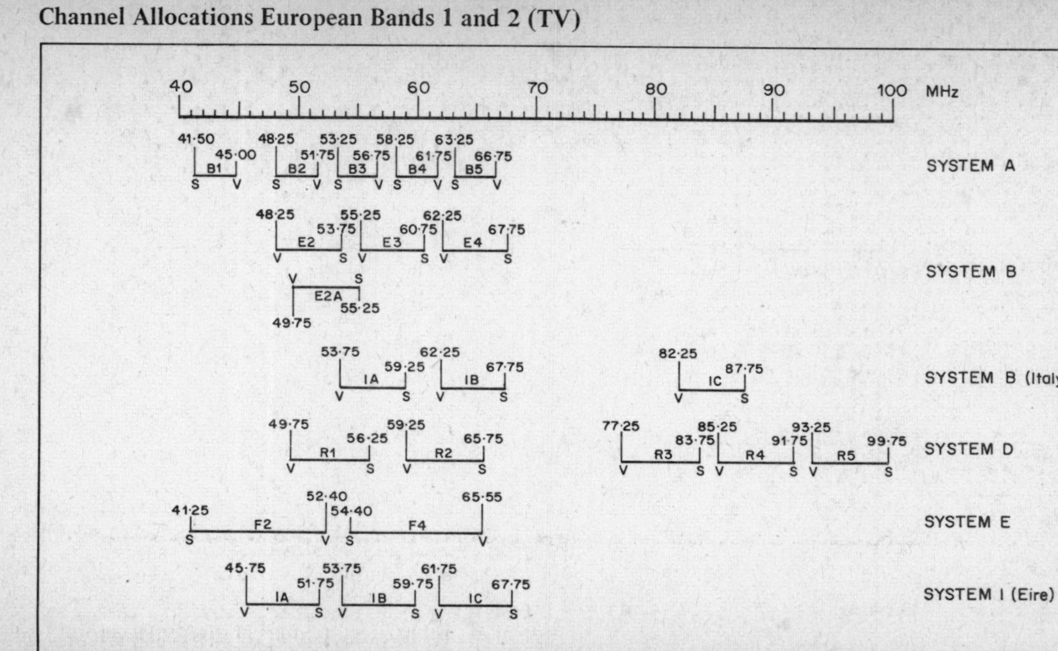

Channel Allocations Band 1 Low Band

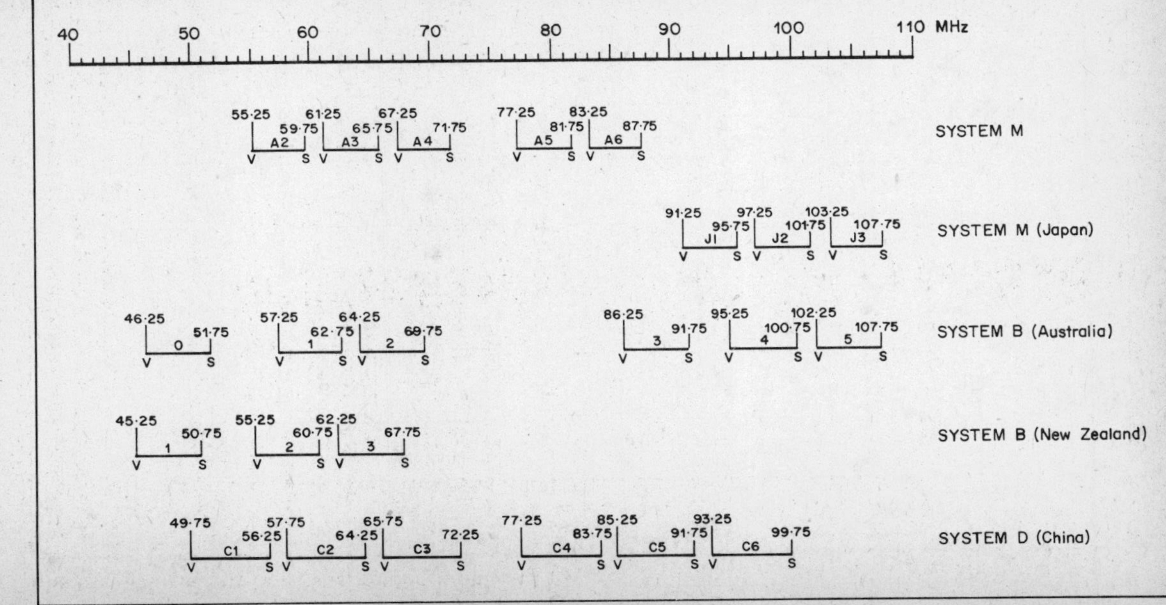

Channel Allocations Band 3

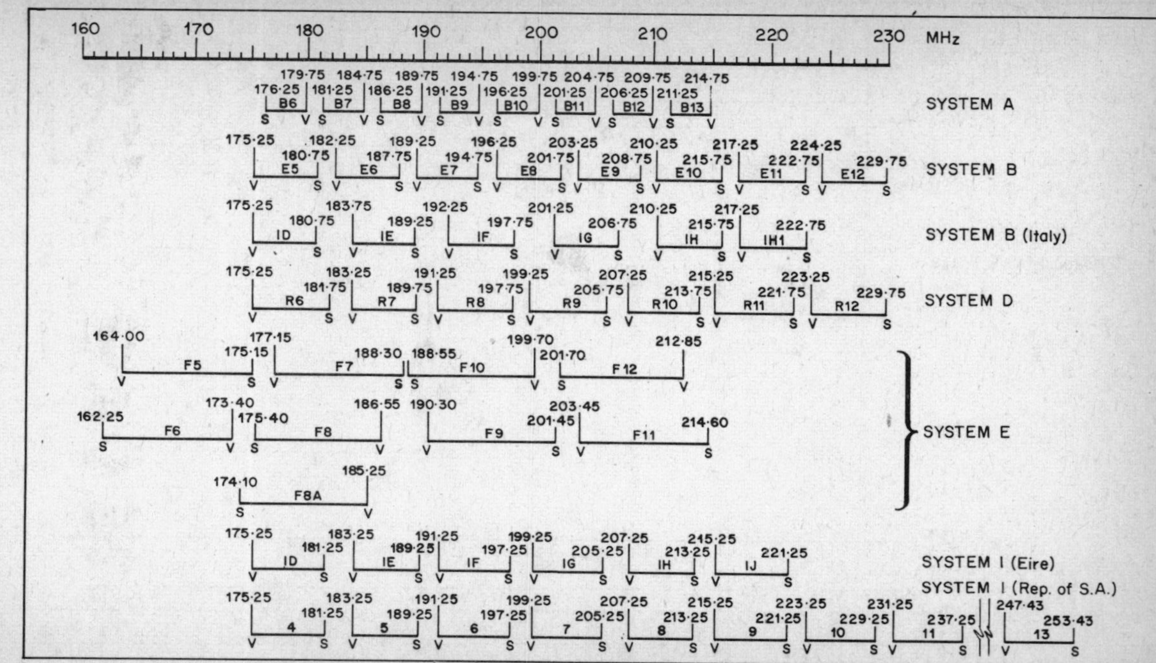

Channel Allocations Band 3 High Band

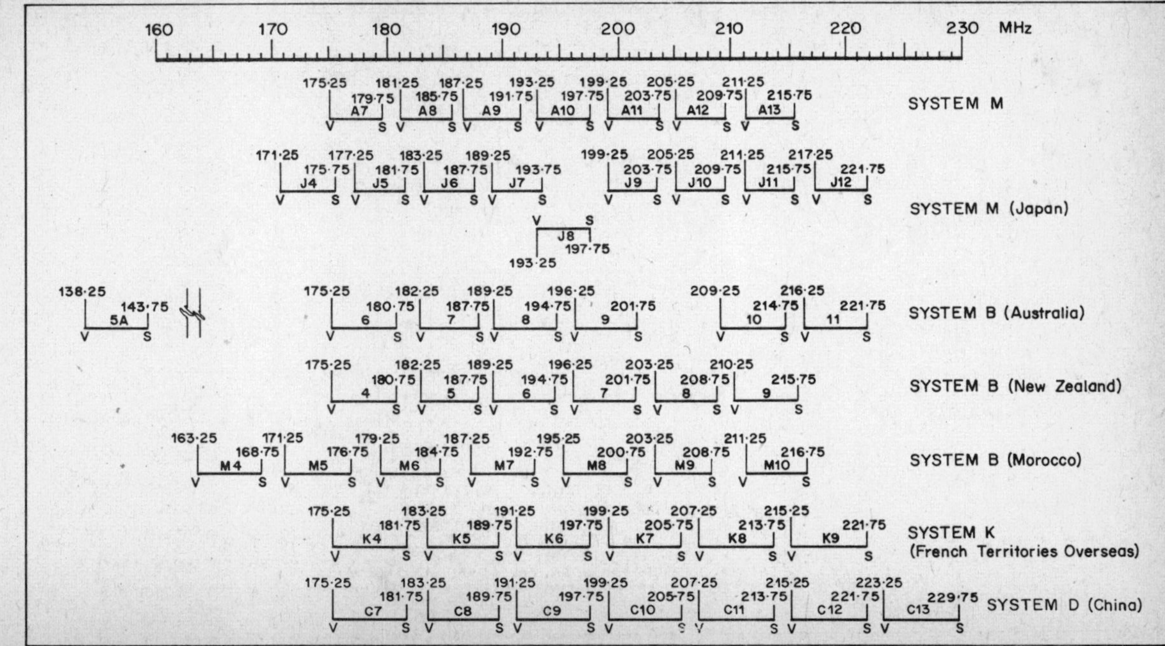

Channel Allocations Bands 4 and 5 – Systems D, G, H, I, K, L
Vision Carrier Frequencies (MHz)
(The Sound Carrier is spaced from the Vision Carrier by the following:
Systems G, H   + 5.5 MHz
Systems I   + 6.0 MHz
Systems D, K, L   + 6.5 MHz)

21	471.25	31	551.25	41	631.25	51	711.25	61	791.25
22	479.25	32	559.25	42	639.25	52	719.25	62	799.25
23	487.25	33	567.25	43	647.25	53	727.25	63	807.25
24	495.25	34	575.25	44	655.25	54	735.25	64	815.25
25	503.25	35	583.25	45	663.25	55	743.25	65	823.25
26	511.25	36	591.25	46	671.25	56	751.25	66	831.25
27	519.25	37	599.25	47	679.25	57	759.25	67	839.25
28	527.25	38	607.25	48	687.25	58	767.25	68	847.25
29	535.25	39	615.25	49	695.25	59	775.25		
30	543.25	40	623.25	50	703.25	60	783.25		

Channel Allocations Bands 4 and 5 – System M
Vision Carrier Frequencies (MHz)
(The Sound Carrier is spaced at +4.5 MHz)

A14 471.25	A28 555.25	A42 639.25	A56 723.25	A70 807.25
A15 477.25	A29 561.25	A43 645.25	A57 729.25	A71 813.25
A16 483.25	A30 567.25	A44 651.25	A58 735.25	A72 819.25
A17 489.25	A31 573.25	A45 657.25	A59 741.25	A73 825.25
A18 495.25	A32 579.25	A46 663.25	A60 747.25	A74 831.25
A19 501.25	A33 585.25	A47 669.25	A61 753.25	A75 837.25
A20 507.25	A34 591.25	A48 675.25	A62 759.25	A76 843.25
A21 513.25	A35 597.25	A49 681.25	A63 765.25	A77 849.25
A22 519.25	A36 603.25	A50 687.25	A64 771.25	A78 855.25
A23 525.25	A37 609.25	A51 693.25	A65 777.25	A79 861.25
A24 531.25	A38 615.25	A52 699.25	A66 783.25	A80 867.25
A25 537.25	A39 621.25	A53 705.25	A67 789.25	A81 873.25
A26 543.25	A40 627.25	A54 711.25	A68 795.25	A82 879.25
A27 549.25	A41 633.25	A55 717.25	A69 801.25	A83 885.25

Channel Allocations Band 5 — System M — Japan only
Vision Carrier Frequencies (MHz)
(The Sound carrier is spaced at +4.5MHz)

J13	471.25	J23	531.25	J33	591.25	J43	651.25	J53	711.25
J14	477.25	J24	537.25	J34	597.25	J44	657.25	J54	717.25
J15	483.25	J25	543.25	J35	603.25	J45	663.25	J55	723.25
J16	489.25	J26	549.25	J36	609.25	J46	669.25	J56	729.25
J17	495.25	J27	555.25	J37	615.25	J47	675.25	J57	735.25
J18	501.25	J28	561.25	J38	621.25	J48	681.25	J58	741.25
J19	507.25	J29	567.25	J39	627.25	J49	687.25	J59	747.25
J20	513.25	J30	573.25	J40	633.25	J50	693.25	J60	753.25
J21	519.25	J31	579.25	J41	639.25	J51	699.25	J61	759.25
J22	525.25	J32	585.25	J42	645.25	J52	705.25	J62	765.25

SATELLITE BROADCASTING FREQUENCY BANDS

Frequency Bands Allocated for Satellite Broadcasting (Region 1 – Europe, Africa, USSR)

620 – 790 MHz	Allocated to the broadcasting service and may be assigned to satellites for Frequency Modulation television subject to administrative and technical agreement.
2.5 – 2.69GHz	Allocated to the broadcasting satellite service and to the fixed and mobile services. Satellite use is restricted to national and regional programmes – subject to agreement as above.
11.7 – 12.5GHz	Allocated to the satellite broadcasting service and to fixed and mobile services.
41 – 43GHz	Allocated to the satellite broadcasting service.
84 – 86GHz	Allocated to the satellite broadcasting service.

Transmission for the European area will be within the 11.7 – 12.5 GHz spectrum

12GHZ European Channel Allocations

	Channels	Polarisation	Orbital Position (Degrees)
Hungary	22 26 30 34 38	1	−1
Poland	1 5 9 13 17	2	−1
East Germany	21 25 29 33 37	2	−1
Czechoslovakia	3 7 11 15 19	2	−1
Bulgaria	4 8 12 16 20	1	−1
Rumania	2 6 10 14 18	1	−1
Denmark	12 16 20	2	5
Denmark (Nordic Area)	24 36	2	5
Finland	2 6 10	2	5
Finland (Nordic Area)	22 26	2	5
Sweden	4 8 34	2	5
Sweden (Nordic Area)	30 40	2	5
Norway	14 18 38	2	5
Norway (Nordic Area)	28 32	2	5
Greece	3 7 11 15 19	1	5
Yugoslavia	21 25 29 33 37 23 27 31 35 39	1	−7
France	1 5 9 13 17	1	−19
West Germany	2 6 10 14 18	2	−19
Italy	24 28 32 36 40	2	−19
Luxembourg	3 7 11 15 19	1	−19
Belgium	21 25 29 33 37	1	−19
Netherlands	23 27 31 35 39	1	−19
Austria	4 8 12 16 20	2	−19
Switzerland	22 26 30 34 38	2	−19
United Kingdom	4 8 12 16 20	1	−31
Ireland	2 6 10 14 18	1	−31
Portugal	3 7 11 15 19	2	−31
Spain	23 27 31 35 39	2	−31
Liechtenstein	3 7 11 15 19	1	−37
Andorra	4 8 12 16 20	2	−37
Monaco	21 25 29 33 27	1	−37
San Marino	1 5 9 13 17	1	−37
Vatican City	23 27 31 35 39	1	−37

Polarisation: 1 Left-hand Circular
2 Right-hand Circular

EIRP: generally range 1½ to 8 MW

Channel Numbers and Assigned Frequencies for the 12GHz Satellite Broadcasting Band (MHz)

1	11727.48	15	11996.00	29	12264.52
2	11746.66	16	12015.18	30	12283.70
3	11765.84	17	12034.36	31	12302.88
4	11785.02	18	12053.54	32	12322.06
5	11804.20	19	12072.72	33	12341.24
6	11823.38	20	12091.90	34	12360.42
7	11842.56	21	12111.08	35	12379.60
8	11861.74	22	12130.26	36	12398.78
9	11880.92	23	12149.44	37	12417.96
10	11900.10	24	12168.62	38	12437.14
11	11919.28	25	12187.80	39	12456.32
12	11938.46	26	12206.98	40	12475.50
13	11957.64	27	12226.16		
14	11976.82	28	12245.34		

Note: UK channels 4, 8, 12, 16 & 20 orbit position 31°W

Proposed Broadcast Satellite Parameters for the Frequency Band 11.7 − 12.5GHz

Type of modulation	fm	
Number of lines	625	
Sound sub-carrier frequency	6 MHz	
Peak-peak deviation	13.3 MHz	
Peak deviation of sound sub-carrier	50 kHz	
Receiver equivalent rectangular noise bandwidth	27 MHz	
Angle of elevation	15°	40°
Luminance signal − unweighted noise for 99% of worst month	34dB	33dB
Sound signal to weighted noise ratio for 99% of worst month	51dB	50dB

PROPAGATION

As mentioned in the introduction the normal service area from
a high powered transmitter at VHF or UHF extends to and just
beyond the optical horizon. Such reception within this area
is termed 'direct' reception and is basically a 'line of sight'
medium — one in which the receiving aerial can 'see' the
transmitting aerial. Since we are interested in the more esoteric
forms of signal propagation and over far greater distances than
relatively near signal sources, direct signal propagation need
not concern us here — at least in the terrestrial sense. Direct
propagation does however have greater implications when
discussing satellite transmission — covered later in this volume
— and with various attempts at airborn transmissions. Television transmissions have already been successfully relayed
from both moored balloons and circling aircraft.

a) Tropospheric

The service area from a television transmitter extends to just
beyond the optical horizon, at that point signals start to fall
off in strength rapidly. Viewers living in such a fringe area will
have noticed that during certain conditions the signal improves
considerably and possibly even noticing interference from
another transmitter, displaying itself as line-pairing or even a
floating picture. Such conditions are related to the prevailing
state of the Troposphere. The Troposphere is that part of the
Earth's atmosphere adjacent to the ground and extending to
about 20,000 feet. When weather conditions become very
settled, as with a slow moving high pressure system (Anti-
Cyclone), Tropospheric propagation is likely to be enhanced
and producing an extended fringe area. Often during the late
Summer/Autumn months such a weather system can become
virtually stationary over Western Europe. The daytime produces
clear, cloudless skies, with a rapid fall in temperature at night.
As the evening approaches following such a warm cloudless
day, the upper air cools, as does the surface temperature, but
at different rates. This produces a boundary or temperature
gradient and hence the formation of an insersion level. Under
such conditions improved Tropospheric propagation will

occur especially if fog forms. Fog is likely to form over the sea and coastal areas and to some extent inland. The improved reception conditions will be noticed as darkness falls, remaining until the following morning, when conditions again deteriorate as the Sun heats the lower Troposphere and disperses the fog. Reception is often favoured along a path parallel with the prevailing Isobar pattern rather than across the pattern. (Note: Isobars are lines joining equal points of pressure.) As the high pressure system moves away another effect — Tropospheric Ducting — often occurs. Signals are ducted along the trailing edge of the system over some considerable distance. The upper air duct can be selective both in frequency and distance, often a distant transmitter is received, whereas a closer transmitter operating on the same frequency is by-passed. If a high pressure system is stationary for some considerable time upper air ducts can form at any time even during the daytime, when conventional Tropospheric signals tend to be less favourably propagated. A fast moving weather system can also produce ducts, either 'pushing' it in front or 'dragging' it on the trailing edge.

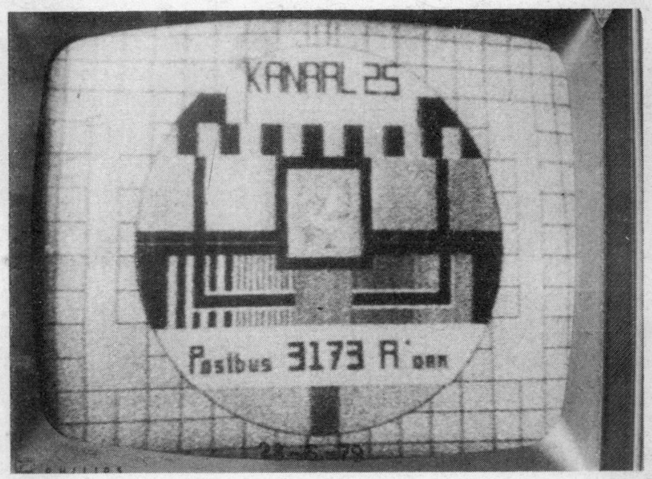

*Tropospheric reception: a 'pirate' TV station
operating in the Netherlands on UHF ch. E25*

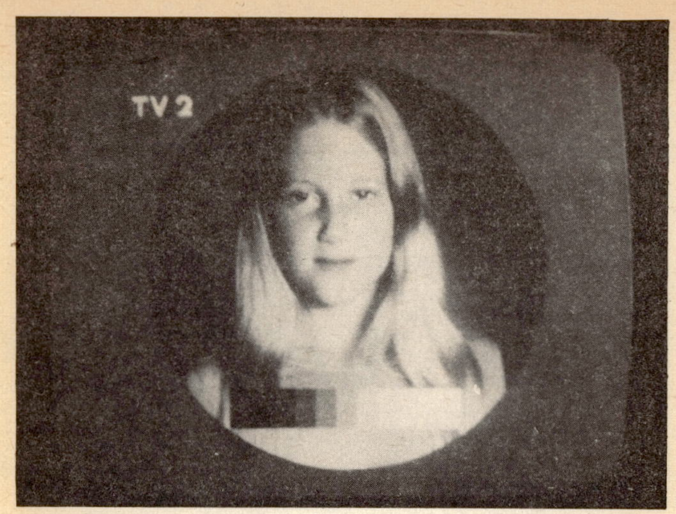

Enhanced Tropospheric propagation: the Goteborg, Sweden ch. E30 transmitter at a distance of 530 miles

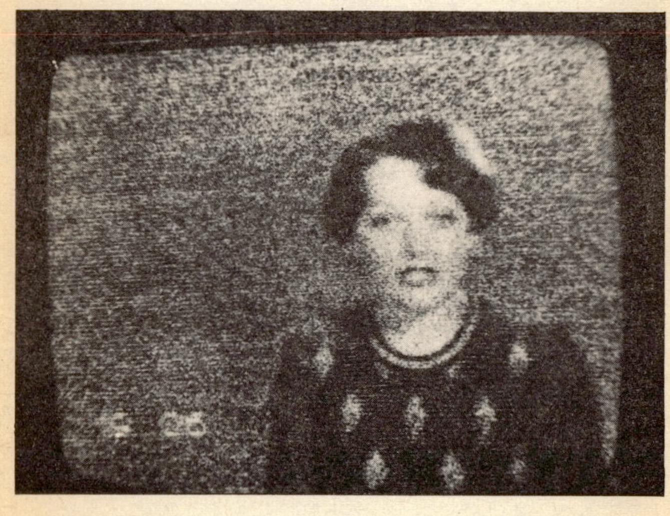

Enhanced Tropospheric propagation: Klaipeda, USSR ch. R29 at 1000 miles

In certain areas of the World, notably the Mediterranean and the Arabian Gulf, ducting conditions can become established for many months of the year and such as to allow regular reception of television signals at entertainment quality, and over distances of several hundreds of miles. Such conditions normally establish during the very hot settled Summer weather and can result in severe interference. The broadcasting authorities in the Arabian Gulf are currently discussing an eventual move to UHF in the hope of reducing the problem to the existing TV services.

The characteristic signal both by conventional Tropospheric propagation and by ducting is a slow fading stable signal. At times selective fading may be noted with the sound frequency fading independently of the vision and vice versa. Frequencies affected cover all those used for television transmission although Band 1 is less favourably propagated than that of Band 3 and UHF. Distances encountered can reach in excess of 1,000 miles for both VHF and UHF signals.

The photographs show typical enhanced Tropospheric reception, one of a slow fading noisy signal over a very long path. At times however both Band 3 and UHF Tropospheric signals can lift to very high levels as is illustrated in the second photograph. Both signals were received at a favourable location near Norwich. East Anglia.

b) Sporadic E
By means of Short Wave radio it is possible to transmit signals to distant countries and indeed around the world itself. Such communication is dependent upon a number of reflecting layers high above the Earth's surface known as the E, F1 and F2 layers. There is a fourth layer — the D layer but this need not concern us here. The E layer lies at an approximate height of seventy-five miles, and under normal conditions reflects Short Wave signals, VHF and UHF signals passing straight through this layer and indeed the F layers and are lost in space. At certain times however, patches of the E layer become intensely ionised and reflect signals back to Earth at frequencies well into the VHF spectrum. During such conditions television transmissions in Bands 1, 2 and very

occasionally Band 3 are capable of being reflected, allowing reception at distances upwards of 500 miles.

The cause of Sporadic E ionisation is not exactly known but various theories have been presented as to its cause. These have included meteor shower ionisation, effects from electrical thunder storms, ionospheric winds producing E layer irregularities, effects from Sunspots and other solar activity and electrical currents flowing within the E layer. No conclusive theory has yet been formulated as to the origin of Sporadic E, indeed there may be more than one cause. Attempts to connect the incidence of Sporadic E with the eleven year Sunspot cycle have proved fruitless — at least on a world wide scale. In early 1980 details of research were published suggesting that Sporadic E may originate as a result of ionisation from meteor ionisation and work continues in this specific field with the hope that the cause of intense Sporadic E ionisation may be established.

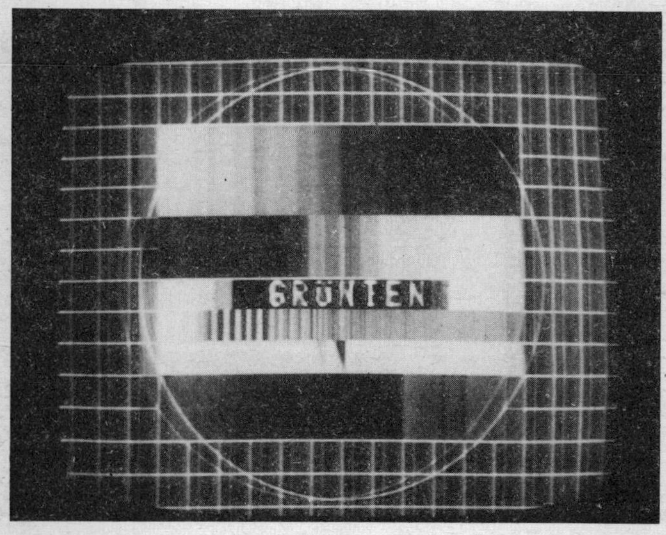

Multipath signal propagation via Sporadic E: the Grunten ch. E2 transmitter in West Germany at 500 miles

Good quality Sporadic E signal: USSR ch. R1 at 1100 miles

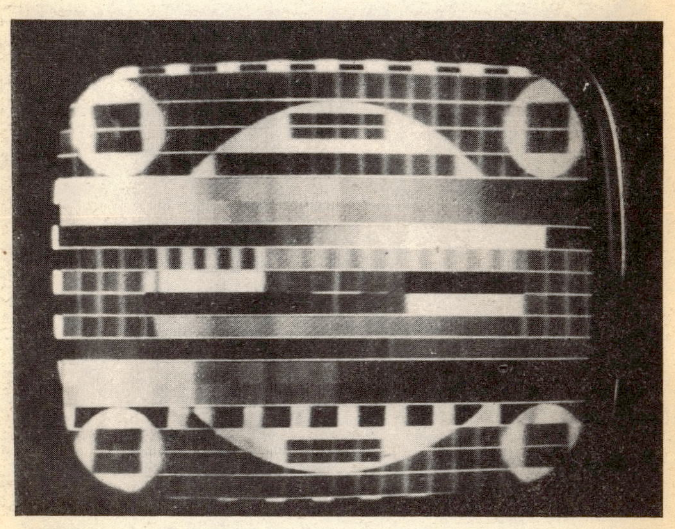

Another example of strong Sporadic E signal reception: USSR ch. R1 at 1100 miles

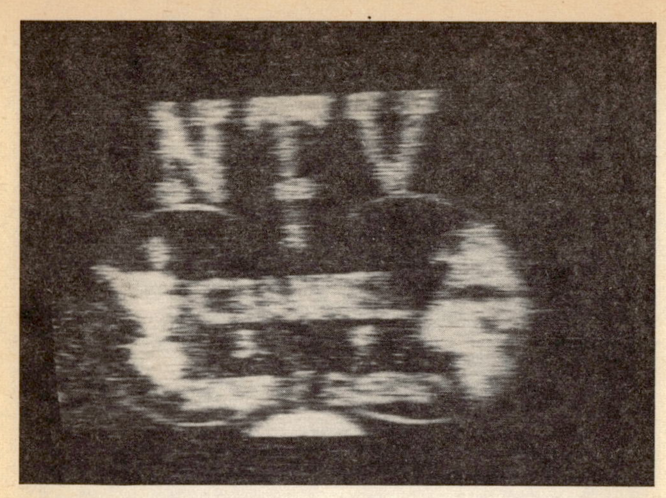

Multiple hop Sporadic E propagation: NTV Sokoto, Nigeria, ch. E3 received in the UK at +3000 miles

An example of Band 3 Sporadic E propagation. The announcer is from RTA Algeria on ch. E5 and received in the Netherlands. Note the additional 'floating' signal in the background on the 5544 test card

Although Sporadic E can occur at any time of the year, the
most active period is during the Summer months, from early
May to September and in some years a small peak of activity
has been noted in mid-Winter. The Sporadic E cloud itself may
be small or widespread and is capable of movement of up to
250 m.p.h. This results in a changing length of signal skip
path, with signals being received from consecutive transmitters,
either close or more distant depending upon the direction of
movements of the cloud. The length of a skip path varies
between 500 – 1,400 miles approximately and at times
double or even triple hop has been known to occur. The signals
as received via Sporadic E can be extremely strong and range in
strength over a few seconds from just detectable to overloading. Polarisation shift can be considerable often being comple-
tely random, although for the longer skip signals polarisation
tends to remain in the original transmitted polarisation and
together with a more stable signal not unlike a characteristic
Tropospheric signal. For the shorter skip signals of up to
1,000 miles or so, signals can be reflected from more than one
part of the Sporadic E layer, resulting in multiple images and
ghosting, with at times phase reversal and other effects. It does
appear that at times part of the Sporadic E reflecting layer may
form in a more vertical plane as often signals are noted
arriving from up to 30° off the correct bearing for the transmitter. From year to year Sporadic E may tend to favour alternative directions and areas such as Sporadic E openings predom-
inating particular stations or countries over that of other regions.

When very short skip Sporadic E is noticed – which will
usually indicate intense ionisation patches within the E Layer,
there is every possibility of E Layer back-scatter. An example
of this phenomenon occured in May 1980 when during an
intense short skip Sporadic E opening the Gort, ch.IB in Eire
was being received in the London area, signals on ch.E3 from
Liege, Belgium were also received but from a Westerly
direction.

Although the frequencies most likely to be propagated are
within Bands 1 and 2 on rare occasions signals have been noted
at the low frequency end of Band 3. If very short skip is seen,

i.e. under 500 miles in Band 1, then there is every possibility that the ionised area will be capable of reflecting a signal of much higher frequency – possibly a Band 3 channel – since a sharp reflection angle (short skip) favours lower frequencies, a shallower reflection angle from the same ionised cloud will favour a high frequency: with a shallower reflection angle a single hop signal will come from a much greater distance. There is a tendency for Sporadic E to occur during spells of thundery weather in the Summer months especially if thunder storms are noted locally.

In the Southern Hemisphere the Sporadic E season also occurs during the Summer period but some 6 months after the Northern Hemisphere, specificly from November onwards. Enthusiasts have noted a link with Trans Equatorial propaga-tion which during the evening period can lead to reception of Korean and Chinese transmitters in Northern Australia.

The duration of Sporadic E signals can range from a few minutes to several hours, often at sustained high signal strengths. For the beginner with little experience of TV-DXing, Sporadic E is recommended for initial experiments, as spec-tacular results can be obtained with the minimum of equip-ment.

c) Meteor Shower/Scatter

Throughout the twenty-four hour period, as the Earth moves through space in orbit around the Sun, it encounters small particles of rock and dust. Such random particles, often no larger than ¼ inch in diameter, enter the various layers that surround the Earth and burn up, appearing as the well known 'Shooting Star'. In addition to the random particles, the Earth often encounters large areas of particles, themselves in orbit around the Sun. Such particles comprise what is known as a Meteor Shower occurring at regular times and periods and are consequently predictable.

When a particle enters the upper atmosphere it burns due to friction. As the particle burns a train of ionisation is produced, generally at the E layer height. If the ionisation density is

sufficiently high it is able to reflect a signal at VHF that
would normally pass through the E layer (except for Sporadic
E as already discussed). Consequently an incident television
signal is capable of being reflected up to distances approaching that of conventional Sporadic E propagation. A signal
reflected by such meteor ionisation can vary in duration from
fractions of a second up to five seconds for really intense
ionised trails, and the signal strengths encountered also vary
considerably from very weak and almost undetectable up to
strong. A particle that enters the atmosphere at an extremely
high speed will burn at a greater height than a slow moving
particle which will burn at a lower height. Consequently a
high velocity particle will tend to give a great skip distance
than that of a slower particle.

The best time for observing random meteors is the early
morning period, when the velocity of the Earth relative to the
velocity of the particles is greatest, although of course such
phenomena can occur at any time of the day or night. The
regular meteor showers can produce extremely good reflections of a fairly long duration due to the number of particles
burning up at E layer height. The more important showers and
approximate dates of arrival are listed below. The exact dates
for any year may be found in various astronomical year books.

Quadrantids	early January
Lyrids	mid April
May Aquarids	early May
Cetids	mid May
Delta Aquarids	end July
Perseids	end July/mid August
Giacobinids	early October
Orionids	mid/late October
Taurids	early November
Leonids	mid November
Geminids	early/mid December
Ursids	mid/late December

Frequencies affected by such meteor scatter propagation tend
to favour Band 1, although at certain times with extremely

intense trails, instances of Band 3 signal reception can occur. Such signals often peak to levels similar to that in Band 1 although generally of shorter duration. The lower frequency channels in Band 3 tend to be more favourably received. It has been noted that if a Band 3 MS 'ping' is observed lower frequency signal 'pings' in Band 1 will follow some seconds later, the latter being of greater intensity and sustained over a longer period. It is therefore often possible to identify a weak Band 3 'ping' if the following Band 1 signal is stronger and of longer duration, this assumes of course that a similar test pattern or programme is seen. It goes without saying that equipment for successful meteor shower work needs to be of high gain and with extremely good synchronisation properties. As signals to some extent will tend to arrive from the E layer at an angle relative to the horizontal, a slight tilt upwards of the receiving aerial may be found to give an improvement, especially if a multi element array is being used with an extremely narrow vertical acceptance angle.

d) F2 Layer

Solar activity has a cycle of approximately eleven years, during which time Sunspot activity rises to a peak and falls again to a low figure. When Sunspot activity increases, so do the reflecting capabilities of the various layers surrounding the Earth, enabling high frequencies to be used with Short Wave communications. The highest reflecting layer, the F2 layer — some 200 miles above the Earth during the Winter daytime, receives Ultra Violet radiation from the Sun, causing ionisation of the gasses within this layer. If Solar activity is sufficiently high the Ionisation Density is such to reflect signals at extremely high frequencies and into the lower VHF spectrum. At this point it may be worth mentioning that the incident signal entering the F2 layer is actually returned by a process of refraction but for simplicity the term reflection is used. At periods of high Solar activity the MUF (Maximum Usable Frequency) rises and the F2 layer is able to reflect signals over considerable distances; a maximum single hop can reach up to 2,500 miles. There is a variation in the height and formation of the F1 and F2 layers depending upon Solar activity, time of the year and whether day or night-time. Similarly there is a variation in the MUF between the

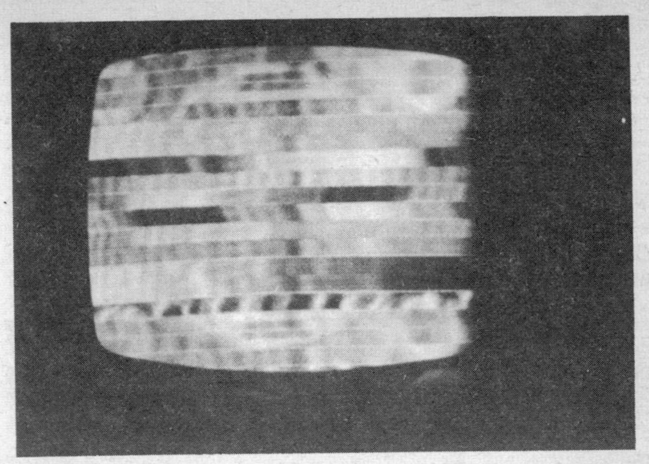

The Russian electronic test pattern received in the UK during February 1980 on ch. R1 via F2 layer propagation

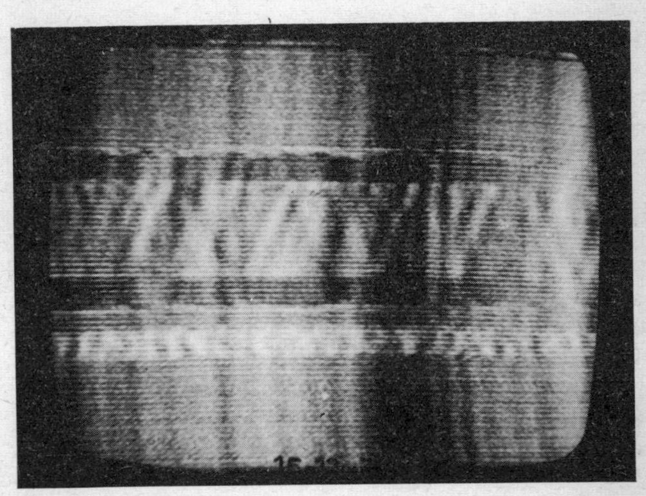

System M 525 line TV signals from CKCW-TV, Moncton, New Brunswick, Canada, received in the Netherlands on ch. A2 via F2 layer propagation in December 1979. Note the characteristic multiple images

*BBC 1 on ch. B1 received in Perth, Australia via intense F2
layer propagation in November 1979*

Summer and Winter, being highest during the Winter day-time,
caused by the less expanded state of the F2 layer, whereas in
the Summer the F2 layer receives more heat and is in a state of
greater expansion, resulting in a lower number of electrons per
unit space.

Generally a signal path running North/South is more favourable
than East/West and in most cases successful VHF propagation
via the F2 layer will necessitate Noon between the transmitting
and receiving points, in order to obtain the required MUF.
Vision signals propagated via the F2 layer tend to suffer severe
multiple images and smearing with phase reversal and
associated effects. Frequencies affected will of course depend
upon Solar activity and can be predicted quite accurately. An
associated effect with high Solar activity is Trans Equatorial
Skip sometimes known as Equatorial Spread F. Towards Sun-
set the two daytime F layers break up and merge to form one
layer at approximately 250 miles high. As the F2 layer breaks
up into small clouds, multiple reflections occur, as signals are
scattered through the cloud region. Such reception occurs

between points either side of the Equator, on a North/South
path, usually limited to a region 40° either side of the
Equator, but at times reaching well into the UK. It is sometimes
possible to obtain signal reflection on an East/West path within
a region 15° North/South of the Equator. Signals, as with conventional F2 layer propagation, suffer with multiple images,
smearing and with a characteristic flutter effect. Recent obser-
vations notes by radio amateurs suggest that Trans Equatorial
Skip may occur during periods when Solar activity is low. Tests
between Japan and Australia confirm that such propagation can
be established within Band 1 and extending to 70 MHz during the
evening period between 20.00 and 24.00 local time, generally the
most favourable period for such conditions to be experienced are
the Equinoxes (ie March/April and September/October).

Both F2 layer and Trans Equatorial Skip (TE) are dependent
upon Solar activity and generally such activity needs to be very
high before propagation at lower VHF occurs. During both the
1956-8 and 1978-80 Sunspot maxima, the reception of Crystal
Palace ch. B1, in the United Kingdom was common throughout the world and at times frequencies up to 55 MHz were
being received across the North Atlantic.

The high level of Sunspot activity that reached a peak in late
1979/early 1980 gave record F2 activity into the VHF bands.
Signals were received on ch. A2 and ch. A3 video (61.25 MHz)
in the UK as were various UK 405 line transmissions noted in the
USA on channels up to ch. B4 video and System B vision signals
to ch. E4. Australian TV signals were received on several
occasions in the UK on ch. A0 and in Australia BBC TV was
relatively commonplace on both channels B1 and B2 with
sound and vision. From a receiving site in the UK a typical
sequence of F2 propagated signals would be distant ch.R1
signals from the USSR at 0800/0830 GMT and suffering
characteristic multiple images and smearing. As the morning
progressed so shorter distance signals would be noted until at
Midday the closest hop signals at 3000 miles (Armenian SSR)
were seen. African stations then appeared and after midday
North American TV transmitters on chs. A2 and rarely ch.A3
would arrive. Conditions during this period were such that

inevitably several stations would be present at any one time making identification impossible. Optimum times for identification seemed to occur at the commencement of skip into a given area when perhaps only one signal would initially be present.

During very active periods a secondary effect of intense F2 propagation is F2 backscatter. An example of this phenomenon can occur from relatively close transmitters when a signal is reflected from the distant F2 layer back to the receiving site but from a seemingly incorrect direction. An example of this has been Czechoslovakian TV transmitters operating on ch. R1 being received in the South UK but the signal direction being the South West rather than the normal Easterly direction. The backscatter signals would be present during F2 openings giving rise to confusion with the multiplicity of signals at both F2 and Sporadic E distances.

e) Aurora

During period of high Solar activity, there is a great possibility of a Solar Flare erupting upon the face of the Sun. When such an eruption occurs, charged particles are emitted from the Sun and spiral towards the Earth, taking about twenty-four hours for the journey. As the particles approach they become influenced by various radiation belts that surround the Earth resulting in a concentration of activity at the Earth's Magnetic Poles (Auroral Zones). The result is an Aurora which often can be seen visually as a display of light and appropriately known in the Northern Hemisphere as the 'Northern Lights'. However the activity is not only seen as visual effects but it also produces ionospheric and magnetic storms within the D, E and F layers, producing a disruption of radio communications.

The Aurora produces a reflecting sheet, which lies in a more vertical plane and it is possible to obtain reflection off this sheet at VHF. As the Auroral sheet lies to the North, it follows that the transmitted signal will be received from a Northerly direction, rather than the true direction of the transmitter, which may be in another direction. It would appear easier to obtain a signal reflection at 90° from the sheet however signal reflection at extremely narrow angles is possible.

There is a tendency for a twenty-seven day re-occurrence from a very large flare due to the rotation of the Sun. A flare situated towards the centre of the Sun will produce considerably more Auroral activity than a flare located on the outer limbs of the Sun. The Equinoxes are the most favoured time for Auroral activity, when either the Sun's North or South Pole tilts towards the Earth, occurring during April and October.

As a general rule aerials should be pointed towards the North varying between North-East and North-West. Frequencies mainly affected are in Bands 1 and 2, although signals have at times been noted up to 200 MHz. On rare occasions, signals have been received across the North Atlantic. Vision signals tend to be poorly propagated via Auroral reflection, generally suffering with severe smearing and distortion. Characteristics of this propagation mode are hum bars on vision accompanied with a rumbling noise on sound. Auroral activity tends to produce two phases of propagation at VHF, an afternoon and a later evening phase have been noted, either phase may be the stronger.

f) Lightning Scatter
Experience has shown that signal reflection may be obtained from the effects of lightning flashes. It would appear that intense localised ionisation occurs in the air around the lightning stroke, this ionisation being of sufficient density to support signal reflection at frequencies through VHF and to the lower UHF spectrum over distances reaching to 500 miles. Such signals are of short duration — resembling meteor scatter — and the lightning flash itself needs to be on or near the transmitter/ receiver path. A storm tracking across the signal path at 90° to the midpoint of the path would provide optimum conditions for such signal reflection. Sheet lightning would undoubtedly provide more positive results than fork lightning. A number of enthusiasts have exploited this medium and confirmed signal reception from transmitters operating at Band 1, 3 and UHF at distances up to 500 miles.

g) Aircraft Scatter

With the rapidly increasing number of enthusiasts in the long distance television field so has been noted the reports of signal reception in the VHF/UHF bands of transmission at distances up to 400 miles, the signals having the characteristic of a typical Meteor Scatter type namely short bursts of a few seconds but of course at too short a distance for MS. It was eventually established that the signals were reflections from high flying aircraft. A number of enthusiasts have erected aerials capable of movement in both bearing and elevation and recorded some success with VHF and UHF signal reflection from high flying aircraft on established flight paths. Such an activity does require a great deal of patience and dedication since the transmitters that are normally received by such a reflective mode are available relatively easily during a slight Tropospheric lift by conventional receiving methods.

h) Field Aligned Irregularities (FAI)

Towards the end of the 1970s and with increasing Sunspot activity, instances were recorded of signal reception at both VHF and UHF but at frequencies that would not be supported by F2/TE propagation. Research has established that high density bubbles of an elongated shape are aligned within the Earth's magnetic field (ie on a North/South path predominently) and giving rise to signal reception within an area ± 2000 miles of the Equator, similar to the evening TE phenomena. The intensity of the 'bubbles' are such that radio amateur signals at 432MHz have been received at distances of 4000 miles. FAI has a characteristic weak and fluttery signal and occurs during the evening period. FAI is associated with high Sunspot activity and is most likely to be received in the Equinoxes (March-May and September-November).

RECEIVER REQUIREMENTS

A television receiver intended for long distance reception has
requirements somewhat more demanding than that of a receiver
intended for 'normal' domestic use. Unlike the domestic receiver
the long distance equivalent has to work with extremes of
signal strength, from a few micro-volts (μV) of a deep fringe or
fading Tropospheric signal, to the extremely high and fluctuating
strength encountered with Sporadic E signals. Receiver gain
therefore has to be of a high order whilst maintaining complete
stability, freedom from overload and associated effects. At
times reception may be sought on an adjacent channel to that
of a high powered local transmitter, which calls for a high
degree of IF selectivity and adjacent channel rejection. The
AGC (Automatic Gain Control) within the IF strip and tuner
should be capable of some adjustment for certain expected
types of signal, whilst retaining sufficient versatility to cope
with the extremes of signal strength and for short bursts of
MS (Meteor Shower/Scatter).

The IF strip itself amplifies the signal from the tuner up to the
order of one volt, at the video detector, prior to feeding the
signal to the video amplifier(s) and it is desirable that this is
accomplished with the minimum of noise possible. The wide
versus narrow video IF bandwidth question will be discussed
later with its relationship to the noise problem but sufficient
to say at this stage that a wide IF bandwidth — necessary for
both sound and video signals — will tend to have a lower gain
and higher noise than that of a narrower IF bandwidth for
video only. In most British receivers only one high gain video
amplifier stage is used — in the case of valves — but it is
common Continental practice to have two stages of video
amplification, the synchronisation information being taken
off after the first stage. With solid state receivers the output
from the video detector(s) is normally passed through a video
driver stage prior to the video amplifier. Synchronisation
information passes to the Synchronising Separator stage,
where the line and field synchronisation pulses are separated
and directed to the appropriate timebase stages. The type and

ability of the synchronising separator stage will vary between
receivers, in some designs various amplification and clipping
stages may be incorporated. Here, as with other stages,
extremely weak signals will be encountered and a first class
performance is required for reception of marginal signals. Flywheel Synchronisation, a circuit that allows for improved
locking within the line timebase must be regarded as an
essential. The requirements for sound reception are somewhat
less demanding; apart from high gain and a low noise figure.

TUNERS

With the growth of the UHF network throughout the United Kingdom so has seen the demise of the VHF tuner being fitted in current production receivers since the 625 line system I transmission is at UHF only. System I is however used in the Irish Republic at both VHF and UHF and receivers destined for this market will have provision for coverage of both frequency spectrums. Before manufacture of the System A receiver ceased the general practice had been to use both transistor and valve tuners and when selecting a receiver for use in long distance reception the valve VHF tuner is to be preferred. The improvement in using transistors over valves at VHF is far less marked than at UHF and there are certain advantages with the use of valves. The bi-polar transistor tends to be rather susceptible to overload and on frequencies closely adjacent to high powered transmissions interference and cross modulation effects may be experienced. The transistor being a low impedance device and with a resulting wider bandwidth property will tend to spread a strong signal over adjacent frequencies, whereas a valve circuit has a more linear response and far less prone to such problems. High performance RF amplifier valves are now found in VHF tuners such as the PC97, PC900, which are capable of excellent gain and low noise figures — the latter point being of great importance. Signals experienced with Tropospheric reception are often of the order of a few μV when arriving at the receiver input and a very low noise amplification is required if we are to avoid the signal being lost in Mixer and IF strip noise.

The VHF tuner may be of several types. The type most often encountered is the Turret Tuner. This contains coil biscuits for each channel required. In the Americas, where only System M signals are likely to be experienced, the tuner can be loaded with coils for channels A2 — A13 inclusive and there are no further problems. In Europe however, with the multiplicity of transmission standards and channel allocations, a considerable number of coils will be required to cover all the channels liable to be received. Another tuner that is in common use operates by means of ganged core assemblies. This tuner is available as either a push button type or with a 'click-stop' mechanism — not unlike a conventional turret tuner to operate. The latter version is the preferred type to use for operational ease and versatility. Unlike the Turret Tuner, the latter version is able to cover all frequencies within each VHF Band which eases the problem of tuning to the various channels presently in use within the European area.

The transistor UHF tuner has undoubtedly ousted the valve UHF tuner on grounds of improved gain and noise figures. In the European area the UHF tuner will always be found with a tuned RF amplifier, although it should be noted that some manufacturers use only three gang tuning — the input to the RF amplifier being untuned (Aperiodic Input). Where possible the four gang tuning types should be used for the improved selectivity which will be useful on channels adjacent to high powered local transmitters.

Unfortunately, in the Americas, RF amplification stages are rarely fitted to UHF tuners, the tuned aerial input being coupled directly into the Mixer stage.

The Varicap tuner — a device which uses variable capacity diodes instead of the more conventional means of tuning — is now increasing in use for both VHF and UHF, although it appears that the selectivity and adjacent channel rejection problems are similar to the conventional transistor tuner, some Varicap UHF tuners are fitted with two tuned RF stages. The German company Telefunken have placed on the market for receiver manufacturers VHF/UHF tuners with FETs fitted in the RF and mixer stages of the VHF section, and indeed for use in areas subject to adjacent channel selectivity problems. Typical figures quoted for an interfering signal two channels away from the desired signal and to give a 1% cross modulation on the desired signal are 100mV as compared to a conventional varicap bipolar tuner with 25mV (Band 3 channels). At the time of writing the FET equipped tuner is becoming available in the UK for amateur useage.

Power supply circuit for Varicap tuners

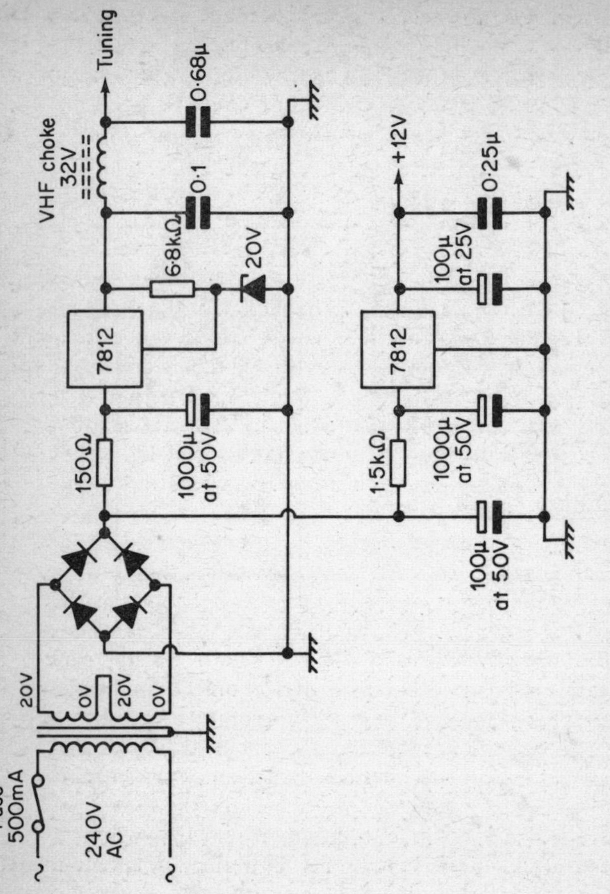

Varicap tuner supply using voltage regulator ICs

The Varicap does however allow a freedom of operation not previously experienced on the conventional VHF tuner. Since it is possible to 'sweep' through all frequencies in Bands 1 and 3, the various channels currently operating in the European area may be received easily, far more so than aligning many individual turret tuner biscuit coils. Varicap tuners for both VHF and UHF are commonly available from component suppliers and an efficient circuit providing power, variable gain, band switching (in the case of the VHF version), coarse and fine tuning is shown. It is possible to modify a conventional VHF tuner, mounted as standard in a receiver, to act as an IF preamplifier. By injecting the IF output of the varicap tuner unit into the modified VHF tuner a vary high gain receiving installation may be obtained.

There is available a transistor device known as an 'Upconverter' which can facilitate receiving VHF signals on a 'UHF only' receiver. This unit is basically a VHF to UHF converter and converts the whole 40 – 250 MHz spectrum to an equivalent range at UHF thus enabling a standard UHF tuner to tune over the complete frequency band at VHF as above. The unit was originally intended for use with communal aerial systems where losses at UHF would be prohibitive. The UHF channels are converted at the head amplifiers to VHF and distributed throughout the system, reconversion back to UHF is then carried out on each individual subscribers receiver via the upconverter.

Since these units are wide band in operation they provide an extremely effective means of covering all channels at VHF including those channels that lie outside the range of most VHF tuners namely chs. R3, 4, 5 and ch. IC Several manufacturers now produce these converter units there being two main types available. One type incorporates a wide band RF amplifier intended for use on systems with low signal levels and the other type omits the RF amplifier and is used on those systems with high level signal distribution. For marginal signal work the former type must of course be used. Great care should be taken to avoid overloading the unit with excessive signal input since successive stages in the receiving chain will tend to exaggerate the condition.

For operational ease the use of push button tuners should be avoided. The turret tuner does allow each channel to be switched to with ease and indeed the slow motion mechanical UHF tuner is to be preferred. With the advent of the Varicap tuner direct channel reading may be taken off of a suitable meter movement, in such cases it is advisable to maintain a table of meter/channel settings.

I.F. STRIPS

The IF strip (Intermediate Frequency Amplifiers) accepts the IF output from the VHF (or UHF) tuner and amplifies the required signal to approximately one volt at the detector prior to feeding the video amplifier(s). It is within this section that the main IF response is determined; the VHF tuner consisting of effectively wide band circuits. In passing, some receivers operating on UHF often utilise the VHF tuner in whole or part as an addition IF amplifier/preamplifier. For effective long distance reception, a minimum of three valve vision IF stages is called for, or four stages if a transistor strip is used. This number of stages is necessary to obtain the required gain/bandwidth performance; bearing in mind the transmitted signal information may have a bandwidth of up to 8MHz (14MHz in the case of System E). Although weak signals produce problems undoubtedly an equal problem is interference from high powered transmitters, thus effectively limiting the use of high gain aerial amplifiers due to various forms of adjacent and co-channel interference. One way of surmounting such a problem is to include various filters and traps within the IF strip, to remove the offending interference from adjacent channels and frequencies. Generally most receivers will have insufficient filtering and additional traps can be added to improve the adjacent channel rejection. If one particular transmitter causes a problem an aerial notch filter can be fitted, but it should be born in mind that during period of enhanced reception, strong signals are likely to appear on otherwise unoccupied channels — producing further problems. Consequently the IF strip should incorporate all the filtering and traps necessary to give good adjacent rejection either side of any channel in use.

It has become a practice within the United Kingdom to make use of suitably modified System A receivers, with their narrower bandwidth video IF strip (3MHz). Such a narrow IF bandwidth results in an increased gain with a considerably lower noise figure, allowing reception of extremely weak signals which on a System B receiver (5MHz video bandwidth) would be marred by noise. This does tend to show a lack of HF detail if the signal becomes extremely strong but the loss is well worthwhile with marginal signals. An additional advantage is that with the multiplicity of transmission standards within Europe and closely adjacent channel allocations, a narrow IF bandwidth receiver is able to tune to each signal, whereas with a System B receiver, adjacent channels tend to float over each other (an example being ch. E2/1). The advantage of a narrow IF bandwidth is only too obvious during a prolonged Sporadic E opening! Where possible the provision of adding wide/narrow bandwidth switching within the IF strip should be investigated. With a narrow video IF bandwidth the sound channel of System B and other similar standards using Intercarrier sound will of course be lost.

With the advent of the single standard System I receiver into the United Kingdom so have IF strips become somewhat less versatile, at least with a view for long distance television usage. Printed circuit IF coils have come into common use with little or no means of adjustment, and the fitting of multi-purpose integrated circuits in IF strips had dispensed with many of the tuned circuits once associated with this part of the television receiver. IF shaping in this latter type of receiver is usually accomplished with an IF preamplifier stage fitted between the tuner's IF output and the input to the main IF strip. Such a stage may comprise one or two transistors and several tuned circuits, if these latter circuits are variable then it is usually possible to 're-shape' the IF response to a sharper curve in the interests of greater selectivity. It may be advisable to fit a second IF preamplifier stage and adjust the latter rather than disturb the receiver's main response, then either the first stage (wide selectivity) or second stage (narrow selectivity) may be switched into circuit. Such a preamplifier stage may be fitted into a receiver having printed circuit IF coils, thus improving both the IF selectivity and overall gain figures. With the increasing use of Surface Acoustic Wave Filters (SAWF) in modern IF strips, adjustment of the response curve will be extremely difficult and in such receivers it may be possible to construct an IF preamplifier stage and obtain the required selectivity performance. A relatively simple 2 stage IF preamplifier is

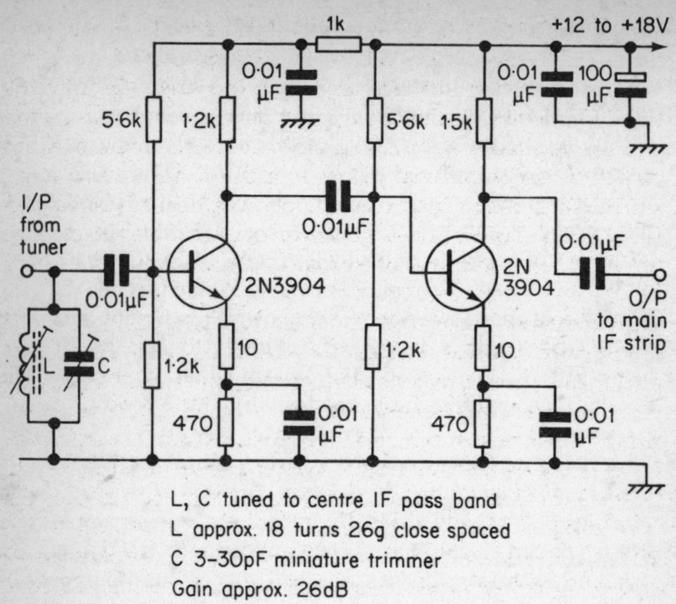

L, C tuned to centre IF pass band
L approx. 18 turns 26g close spaced
C 3–30pF miniature trimmer
Gain approx. 26dB

Preamplifier for IF strip gain improvement

shown that gives a useful gain of approximately 26dB. Being a wideband circuit the bandpass characteristic is determined by the tuned circuit L and C at the circuit input. A further tuned circuit can be added at the output prior to connection to the main receiver's IF strip.

In modern receivers IF circuitry may be contained in one or more discrete modules and in certain of the Philips range these are designated as an IF selectivity unit and an IF gain board. The former contains several tuneable coils and a single transistor stage where-as the latter features the bulk of the IF gain and with minimal coil adjustment. Since the IF selectivity unit adjusts the overall shape of the IF bandpass it follows that the fitting of this module into any receiver of similar IF will enable modification of the IF bandpass by careful adjustment of the module's coils. It is possible to fit a module of this type in series with the IF input feed from the receiver's tuner(s) to the main IF strip and 'peak' up the module to provide a restricted bandpass and hence improved selectivity. In certain circumstances two such modules have been inserted in series to give sharp selectivity. It is then possible to either insert into circuit for narrow selectivity or to by-pass for wide selectivity with a simple slide switch or pin diode switching. The Philips U800 Selectivity Module is available from television spares/surplus dealers or from Hugh Cocks Television Services.

The reduced IF bandwidth does give considerable help with reception of marginal satellite originated TV signals, the trade-off for an improved signal/noise figure against that of a reduced video bandwidth. Experience has shown that a signal not visible on a receiver using an IF bandwidth of 5MHz, can be resolved using a receiver with its bandwidth restricted to 2.5MHz.

In the Americas with only one standard likely to be received — that of System M — particular attention need only be paid to the adjacent channel protection figures. Often a simple IF preamplifier stage can be fitted tuned to the centre of the IF passband and using only two notch filters — that of adjacent channel video (LF) and adjacent channel audio (HF).

With the three main sound/vision spacings of 5.5, 6, 6.5 MHz (Systems B/G, I, D respectively), so efficient coverage of the matching sound signals — when in a wide IF bandpass mode and using a single standard receiver — is extremely difficult. Several experienced enthusiasts have successfully modified their receivers (that use an IC type sound IF) by switched varicap diode tuning of the sound take-off and IC detector coils. Since only two tuned circuits are involved, 3 position preset switching can be incorporated relatively easily although correct may prove difficult without relevant test equipment.

VIDEO AMPLIFIER

The IF strip terminates in the video detector which rectifies the incoming IF signal prior to the video amplifier(s). Although circuitry here is conventional, again in Europe with reception of either positive or negative going video, switching has to be incorporated. Many European television receivers incorporate such switching together with the necessary IF adjustments for both video and sound (and timebases where necessary). Consequently little problem should be experienced with receiving either video standard on such receivers. However for the single standard receiver, a second video detector diode should be fitted — with low loss switching and wiring — enabling either standard to be resolved. Generally some adjustment to video amplifier bias should be made when switching can be incorporated upon the same switching assembly as the detector(s). Assuming a positive going video receiver is being modified to switch to negative video, a reduction in cathode bias on the video amplifier is all that is usually required. Comparison should be made with circuits of commercial receivers that incorporate video switching, from which source similar circuit information can be extracted.

Three typical video detector circuits are illustrated that give provision for positive or negative video signals to be selected. The valve circuit may be found in the more elderly receivers and generally was superseded by the semiconductor circuit shown in figure (b). These circuits can be incorporated in a single standard receiver to provide switching between 2 video standards but extreme care must be taken to minimise connecting lead lengths in the interests of gain and stability. In many solid state receivers an emitter follower stage follows the video detector and this can be modified relatively easily to provide positive and negative switching (+ video at collector, — video at emitter). A small switch may be used to select either video sense and since the video is at low impedence, connecting leads may be permitted to be slightly longer than when modifications are made directly onto the detector. Figure (c) gives a typical video phase splitter which can be

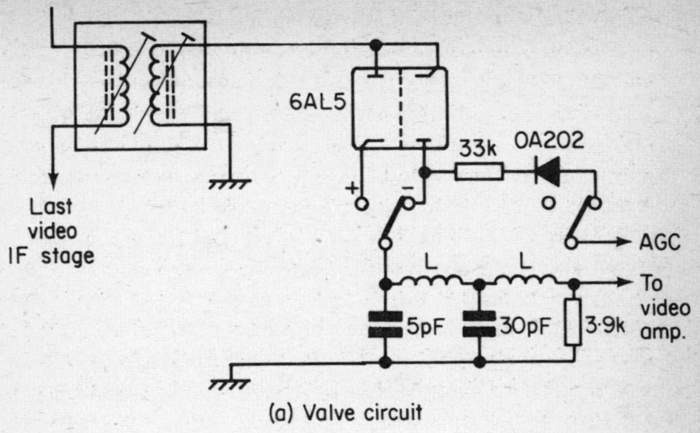

(a) Valve circuit

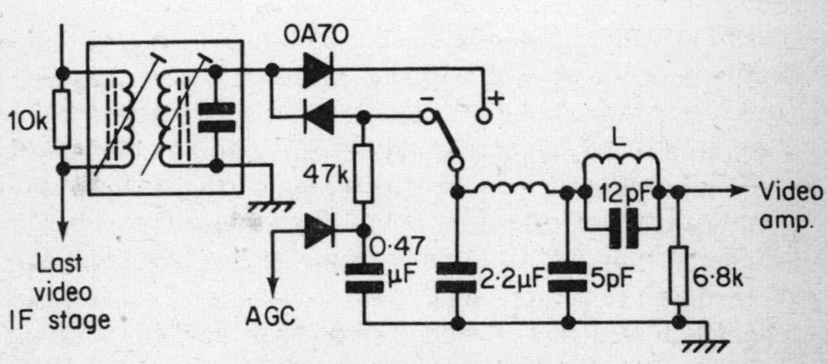

(b) Semiconductor circuit

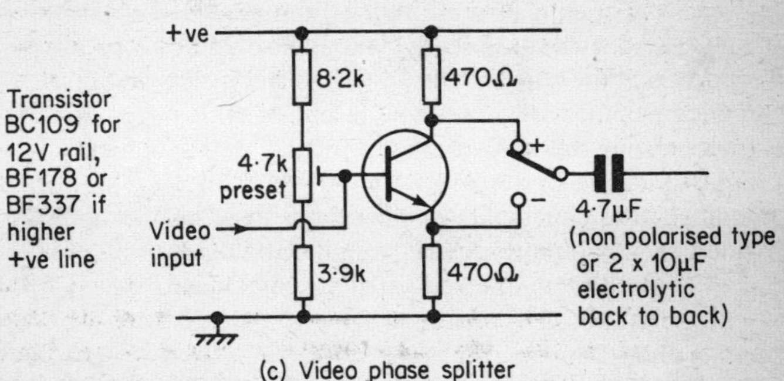

(c) Video phase splitter

*Typical detector circuits for switching positive or negative
video modulation*

fitted into a single standard IF strip or in the case of a
receiver already fitted with an emitter follower between
video detector/video amplifier this can be modified to
values similar to figure (c). An AGC feed may be found in
certain receivers from the emitter of the video emitter
follower stage. This should be left as the operation of the AGC
circuits will be unaltered by the video phase splitting modifications. Only the very experienced should attempt modifications of a receiver using ICs incorporating video detection.

From the video amplifier(s), one output is taken as drive to the
cathode ray tube and another to the Synchronising Separator.
This stage extracts the various line and field pulses necessary
to synchronise the appropriate timebases. Various types of
synchronising separators will be found, one particularly
efficient version is the noise gated type which is suitable for
deep fringe work. Various clipping and amplification stages
may be found indicating its suitability for weak signal work.
Of the utmost importance is a receiver incorporating Flywheel
Synchronisation which is essential for any long distance television work. The synchronisation of the timebases must be
accurate and correct, as certain signals such as Meteor Shower
(MS) produce very short bursts of signal which necessitates
immediate locking if the signal is to be identified. Often it is
possible to improve the filtering and performance of the
synchronisation separator stage for such weak signal work. One
new system coming into use which improves field sync. locking
makes use of a count down system. Basically line sync. pulses
are divided down developing a field sync. signal to give correct
field lock immediately on signal reception. This system obviates
the need for a field hold and field oscillator stage and early
reports indicate its suitability for extremely weak signal work.
Generally the sync. separator stage has little scope for modification and apart from experimenting with increased filtering
and increased values of sync. coupling components the
section should be left unaltered.

Often derived from the sync. separator stage is a negative
going voltage which after filtering is applied as IF amplifier
bias (in the case of valves), reducing the gain on strong signals

and vice versa. Such a system is known as Mean Level AGC
(Automatic Gain Control); the voltage being dependent upon
the average brightness of the video signal. Other systems derive
AGC voltage from the line output stage known as Line Gated
AGC and Sync. Pulse — cancelled AGC. Solid state receivers
differ somewhat with AGC circuitry and often incorporate
an AGC amplifier stage to derive controlled stages, reference
signal levels again may be sampled on sync. pulse intervals
or from the video stages.

The AGC voltage to the RF amplifier is usually delayed in an
endeavour to maintain best signal/noise figure possible, in
certain single standard System A receivers of the 1960s an
adjustable delay could be selected for optimum picture quality.
Often the RF amplifier within the VHF tuner has provision to
be switched to either the delayed AGC or to chassis, the latter
position enabling an improved signal/noise performance. The
time constant of the AGC line is normally fixed but with an
advanced receiver a switched time constant may prove invaluable, especially with Meteor Shower (MS) signals, which by
their nature are extremely short in duration and with a wide
range of signal strengths. One disadvantage with gated AGC
systems occurs when a signal suffers from a severe ghosting or
multiple images. A negative ghost can cancel the synchronisation thus reducing the efficiency of the AGC system.

The timebases themselves are conventional but it is necessary
for them to remain stable for long periods. Again in Europe
there are several line standards which need to be catered for.
Fortunately information for such adjustments on a System A
receiver that may be required can be obtained easily by
scrutinising multiple standard television receiver circuits and
generally any modifications can be accomplished quite easily.
It is a general practice to feed the field oscillator from the
boost HT line in the interests of stability. There may be a
tendency with some receivers on fast fading signals for the
line oscillator to fall out of synchronisation for short periods
during such deep fades. At such times the boost HT may alter
somewhat, causing the field oscillator to alter in frequency
and producing a roll. It may prove useful to connect the field

oscillator, suitably decoupled, to the receiver main HT line.

Power supplies are conventional but it should be borne in mind that within the United Kingdom it is common practice to find live chassis technique (AD/DC). Consequently great care must be taken whilst operating or adjusting receivers with protective covers removed. Death is so permanent!

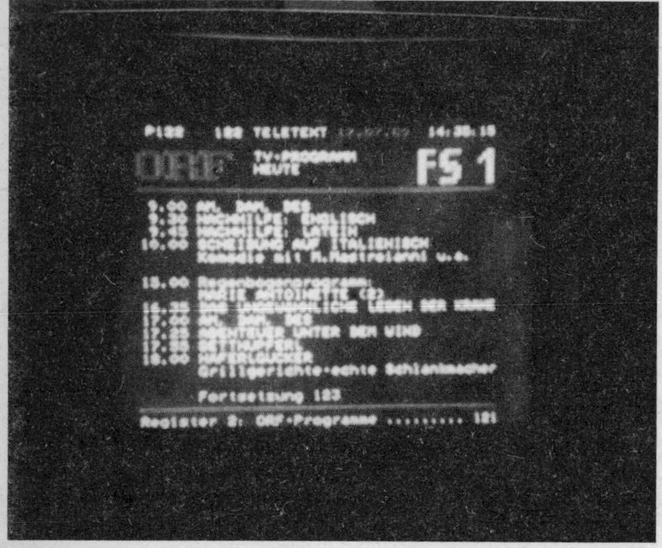

Teletext information is being carried by several broadcasting networks as this off–air DX shot illustrates from the ORF (Austrian 1st TV ch.)

COLOUR RECEPTION

Colour broadcasts are now being transmitted regularly in many countries. It is possible to receive such transmissions over long distances for both Tropospheric and Sporadic E propagation. By the nature of the transmission itself successful colour reception demands a higher standard of receiver efficiency if satisfactory colour signals are to be displayed. A number of colour systems are in use. PAL, SECAM and NTSC. The first two types are used in Europe, PAL by Western Europe excluding France, and SECAM in Eastern Europe and France. NTSC is used in the Americas and areas with the FCC 525 line system.

Unlike monochrome reception where video bandwidth can be sacrificed in the interested of an improved signal/noise figure and IF selectivity, it is essential that the IF strip retains the full bandwidth and response as laid down in the manufacturer's specification. Similarly it is essential that the aerial in use maintains a level response throughout the design bandwidth since the colour subcarrier lies at the high frequency end of the channel, if the aerial has been peaked to the vision frequency with an appropriate fall-off in performance towards the HF end difficulty may be experienced in locking weak long distance signals. Experience has shown that it is possible to reduce IF bandwidth to a degree that will prevent adjacent channel interference and retain normal colour operation. Such modification has been carried out by an experienced technician in the Arabian Gulf area to reduce interference problems and with excellent results. It is advisable however to avoid undue modification to any colour receiver's IF strip and associated circuitry unless one is fully experienced in television engineering, the success of any such work will depend on the versatility of the receiver's design to accept additional alteration in a critical part of the circuit.

The chrominance/luminance ratio should therefore be maintained, preferably with a variation not exceeding 6dB, considering that the video bandwidth may be adversely affected in propagation.

The signal levels at which colour will 'lock' depend mainly on the efficiency of the colour killer circuitry, but it should be possible to obtain a colour lock simultaneously with normal synchronisation. Disabling the colour killer does help to receive weak signals in colour. This can normally be done by shorting out two appropriate test points or shorting a particular test point to chassis — the manual should be consulted for such adjustment. If the colour killer is left disabled colour noise will be displayed when receiving monochrome pictures. It is suggested that some arrangement be made to switch the colour killer 'in or out' of operation in the method as indicated. Experience has shown that colour signals by Sporadic E and Tropospheric propagation can be received, although the former mode is more difficult due to the continually fluctuating nature of the signal. Both SECAM and PAL colour signals have been received in the United Kingdom, although it appears easier to obtain a colour lock with the PAL system, as it has a simpler ident signal enabling lower signal working.

Noise on a SECAM signal tends to appear as streaking whereas noise with PAL has a finer speckled appearance.

When operating a PAL colour receiver in the United Kingdom (6MHz sound/vision spacing) with West European PAL signals (5.5MHz sound/vision spacing), patterning on the chrominance information has been noticed with some receivers, due to insufficient rejection of the 5.5MHz sound intercarrier. In such a case extra IF sound traps should be incorporated.

In general terms for satisfactory colour reception, an input signal level some 8–10 dB higher than for monochrome reception is necessary for acceptable domestic viewing.

VCR OPERATION

With increasing use of VCR equipment and availability of VCR
tapes at competitive prices there is every reason to exploit the
medium to improve results in the DX-TV field. The main
advantage is of course the replay facility and distant signals
can be recorded and played back to review specific sightings
be they of exotic reception or of unidentified test patterns
or captions. In the latter context the 'freeze frame' facility
that is available on VCR machines will enable an unknown
pattern to be examined frame by frame and in the hope of
noting some form of identification. A signal incoming
directly that suffers poor synchronisation can often be
improved by recording onto tape and playing back through
the VCR system. If an enthusiast has access to a 2nd
compatible VCR machine it is possible to dub off sections of
wanted video onto a 2nd cassette and to build up a complete
library of test card and other DX video information. Such
recordings are invaluable when demonstrating the art of TV-
DXing to the layman, or for playing back for one's own
satisfaction during the cheerless days of Winter!

IMPORTED DUAL STANDARD TELEVISION RECEIVERS

For the non-technical TV enthusiast the attraction of the
imported dual standard receiver for DXing use is considerable.
This type of receiver, often with a relatively small screen,
incorporates both VHF and UHF tuners, usually varicap tuned
– and features intercarrier sound IF switching from the System I
UK standard to System B/G West European standard, being
6MHz and 5.5MHz respectively. One problem that can arise,
particularly if sited near to a Band 1 transmitter is that the
latter signal – if strong – can 'spread' over a considerable
part of the useable spectrum and prevent operation on one or
more potential DX channels. The 'spread' is caused by the
relatively wide response of the receiver's IF strip, and short
of modification internally the only method of reducing the
problem is with the use of notch filters to remove the offending
local transmissions. The problem may occur if for example
strong signals are received on such a receiver via Sporadic E
propagated signals on the adjacent channels E2 and R1. Due
to the IF characteristics the 2 signals will tend to superimpose
onto each other and with no means of increasing IF
selectivity there is little that can be done. It may be possible
to ease the problem somewhat with the use of a variable tuned
notch filter and with the multiplicity of channels within the
Band 1 spectrum a varicap tuned filter will prove operationally
efficient. The design of such a filter is given within the
'Interference' section later in this volume.

SATELLITE TELEVISION TRANSMISSIONS

At the time of writing (mid 1980) several periods of direct TV reception on domestic receiving installations from satellites in synchronous equatorial orbit have taken place. In 1975/6 the ATS-6 'Site' experiment involving instructional television programming for the Indian sub-continent from its orbit at 35°E was most successful. Transmissions were at 860MHz using wideband FM video modulation and with two sound channels. Despite the main signal beam being fixed on central India sufficient radiation 'off-beam' allowed several enthusiasts in Western Europe to receive signals of reasonable quality and on home constructed equipment. The USSR has at least one direct broadcasting satellite (Statsionar T) operating at 714-722 MHz and providing television for Northern Russia. Despite the 99° East position of the satellite, sufficient

An artist's impression of the Communications Technology Satellite (CTS) in synchronous Equatorial orbit at some 22,300 miles above the Earth's surface.
(courtesy of British Aerospace)

Canadian engineers adjust a domestic 12 GHz dish and
head assembly in use for the CTS experiment

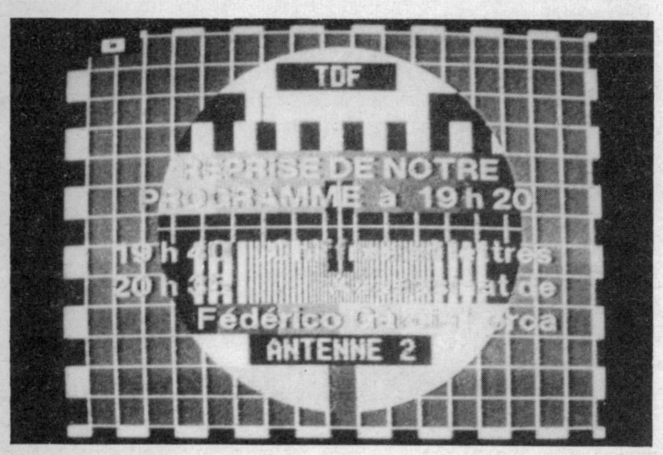

Experimental transmissions from the OTS-2 satellite at
11.6 GHz and using home constructed equipment

radiation due to characteristics of the transmitting aerial exists to enable experimenters as far distant as South Africa to resolve weak signals on relatively simple domestic equipment when using high gain receiving aerials. The future of direct satellite broadcasting, certainly for Europe will lie in the 12GHz band. The Canadian CTS transmissions and later BSE transmissions for the Japanese islands have demonstrated that direct transmissions are capable of giving excellent results. Towards the end of 1978 the experimental OTS satellite was successfully orbited and commenced a series of low power test transmissions. It is to the credit and patience of at least one TV enthusiast that he successfully constructed an 11.6GHz receiver to display good quality video.

The 12GHz head and dish system are such that extreme precision is necessary in its construction. It is likely therefore that the enthusiast will be unable to produce his own head system. Furthermore it seems that there will be available mass produced 12GHz head units when direct satellite broadcasting to Europe commences. The enthusiast can however construct much of the receiving equipment and proven circuitry is included within this section. The 12GHz head unit will give wideband conversion to the 1st. IF, this within the UHF band and the scope of conventional domestic UHF tuners. It is therefore possible to feed the IF output of a suitably tuned UHF tuner into the Phase Lock Loop demodulation circuit as shown. The AM video output can then be fed into a conventional video monitor or remodulated back to VHF (or UHF) and fed into the normal domestic TV receiver.

As already mentioned several enthusiasts received pictures from ATS-6 and examples are shown together with details of equipment used. A 5 foot diameter dish, fabricated in 12 sectors of ¼ inch expanded aluminium mesh and with a 22½ inch focal length. The driven element is a tuned circular quad element fed for vertical polarisation, spaced at ¼ wave from a ½ wave diameter copper disc reflector. A 2 stage transistor head amp (18dB gain, 3dB noise) is fitted to rear of the reflector. Coaxial cable feeds the UHF signal to the receiving equipment comprising a standard varicap UHF tuner into

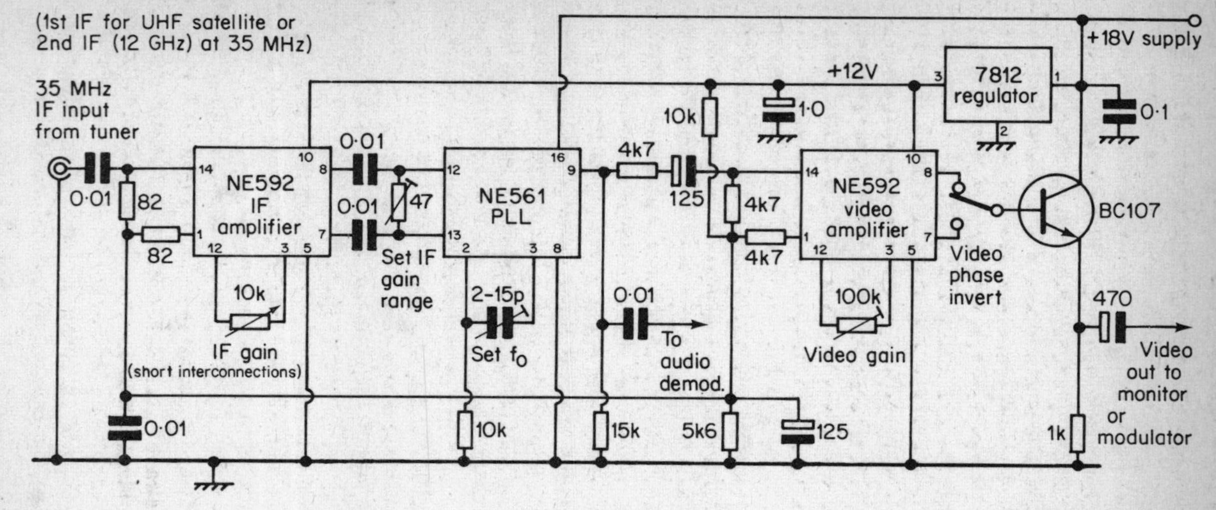

Experimental IF strip and FM video demodulator for satellite reception

A home constructed 5 foot diameter parabolic dish used for satellite reception experiments

Transmissions received from the ATS-6 satellite at 860 MHz in 1976 at Sheffield, England

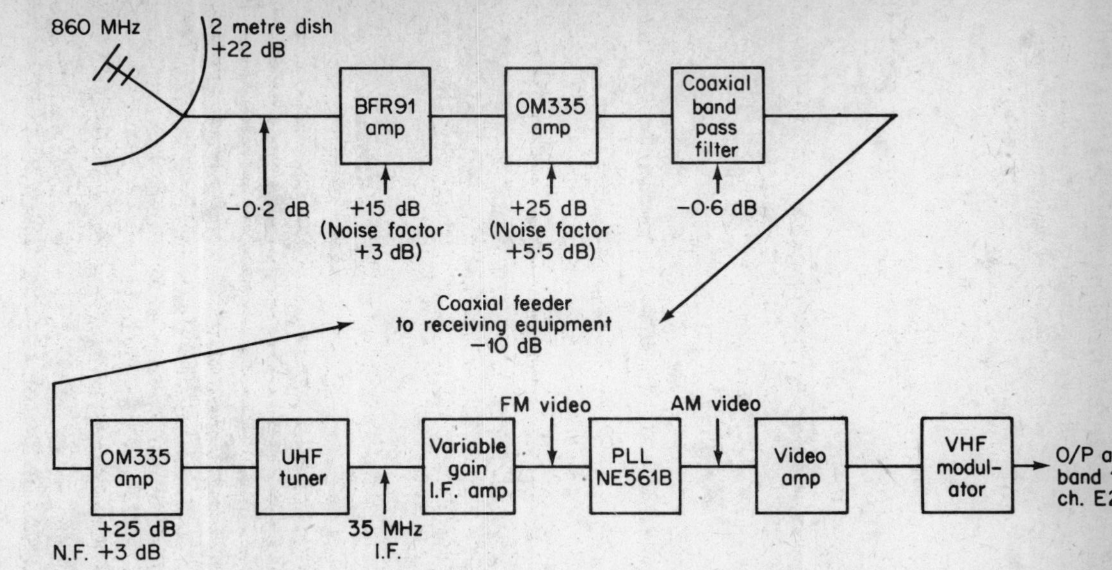

The Peter Jansen 860 MHz satellite receiving unit

the circuit similar to that shown but with an additional PLL.
One loop with a 3MHz lock range feeds video to a monitor,
the 2nd. loop with only 500KHz lock range explores sync.

*A 2 metre diameter aerial used for ATS-6 reception in Holland.
Note the plane polarised (vertical) head assembly*

tips and feeds sync. information to the same monitor. An improvement of 3dB noise bandwidth was obtained and enhanced sync. locking resulted by using the 2 independent

An 8 foot diameter dish used for satellite reception
experiments in Northern England

High quality video from Russian satellite Gorizont 2 at 3675 MHz, and displayed on home constructed equipment in the UK

loops. The precision required at 12GHz. is such that no details will be given for head construction. For the enthusiast that considers himself qualified to attempt such construction he would be advised to study the RSGB publication 'VHF-UHF Manual' by D.S. Evans and G.R. Jessop.

Any form of satellite reception on a line of sight path will involve a free space path loss over the circuit — this signal loss will vary according to the frequency in use and the path length involved. Tables are available in communication handbooks detailing such path losses. A further problem occurs in the avoidance of polarisation shift — this would happen to a plane polarised signal over a long path — the signal is transmitted circularly polarised. For optimum reception results a helical aerial is required to match the incoming wavefront which may be either clockwise or anti-clockwise polarised. It is possible to use crossed yagis for this purpose, reversal of polarisation screw being achieved by switching specific lengths of cable into the aerial feeders. Experience suggests that the polarisation rotation problem is greatly reduced at frequencies above 1GHz (1,000 MHz). Generally the use of a dish parabola is advised above 1GHz whereas below this figure the gain of a parabola is such that stacked yagis will give a greater gain and cleaner polar diagram.

The construction of an accurate dish should present little problem to the dedicated enthusiast. Information and dimensions are given in a number of amatuer radio handbooks devoted to microwave activities. A dish for UHF reception can be constructed using ½ inch mesh chicken netting for maximum efficiency this dimension should not be exceeded. Fibre glass and aluminium foil offer other possibilities and copper pipe can be used in the construction of circular polarisation feed elements.

The future of direct satellite transmission is such that UHF will be adopted for covering large areas from one transmitter where as higher frequencies will be used for covering smaller areas and countries such as in Europe. Such a UHF transmission would have a main beamwidth of between $7^o - 10^o$ in covering large areas — such as India; for transmission to individual countries in Europe a beamwidth between $0.5^o - 1.5^o$ will be employed and operating within the selected band $11.7 - 12.5$GHz $(11,700 - 12,500$ MHz). Due to power considerations wide band frequency modulation is the most likely system to be adopted having a peak to peak deviation of approximately 16MHz and a channel bandwidth of 24 MHz.

Receiving aerials will vary between the traditional parabolic dish and newly developed planar arrays printed onto copper clad laminates. For sufficient gain the aperture of the dish system will be approximately one metre and the aperture of the laminated array will need to be approaching 1.6m square for an equivalent gain figure. Due to the high losses at such frequencies the incoming 12GHz signals will be downconverted adjacent to the aerial system with further conversion at the receiver to video and remodulated to either VHF or UHF AM video. NHK, Tokyo have perfected a design that gives direct to AM video downconverted to UHF or VHF at the head itself

A Philips experimental 12 GHz receiving dish and head converter (1.2m diameter)

with a system noise figure of below 4.5dB. It is anticipated that this design will become available for mass production of low cost domestic satellite receivers.

Currently considerable research is taking place with a view to evolving new reception techniques at these frequencies which will inevitably bring improvements in aerials and conversion methods.

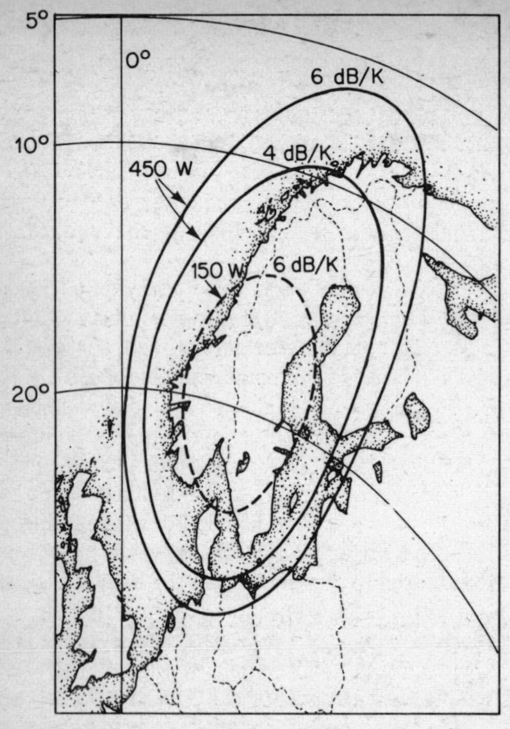

Projected example of 'footprint' showing areas of satellite TV coverage using on-board high (450w) and low (150w) power transmitters. The 'dB/K' are figures of merit for a given receiver system. A 6 dB/K receiver system can be used over a wider area (or with lower field strengths) whereas a 4 dB/K has a more restricted performance and is used with higher field strengths

AERIALS

The aerial is one of the most important sections of the receiving chain, as the signal voltage conveyed to the aerial input of the television receiver will determine the eventual quality of the image displayed upon the screen. Consequently every effort must be made to obtain the highest possible signal voltage.

The use of a wideband Band 1 aerial is advocated in later pages for successful TV-DXing and the UK enthusiast has the option of either constructing his own array or purchasing from several commercial types that have been made available specificly for the 'TV-DXer'. For Band 3 and higher commercial wide band arrays are commonly available and certainly due to the more exacting aerial requirements it is not advisable to construct one's own aerials for such frequencies. It is unlikely that a home constructed aerial would give the same degree of matching and performance over these wide bandwidths as would be encountered on a correctly designed array for Band 3 and above. In the Americas with the large number of transmitters in operation, wideband arrays are extensively used, it being not uncommon to encounter a single aerial covering ch. A2 − A83 inclusive. Generally the enthusiast within this area will operate a wide band array covering VHF and another wide band array for UHF, unlike his European counterpart that uses separate aerials for each band. Certainly an improved performance must be expected when single band aerials are used and where possible this course is strongly advised.

The most simple aerial encountered for television reception is the half wave dipole. This comprises a single rod cut to the appropriate operating frequency. Such an aerial, vertically mounted, will be capable of receiving vertically polarised transmissions through a complete 360° arc and is termed as having an omnidirectional receiving pattern. A half wave dipole mounted horizontally will receive horizontally polarised transmissions at maximum strength when the dipole is broadside to the signal path. Consequently there will be two main receiving lobes for such an aerial, at 180° to each other and if plotted on a circular scale will show a receiving pattern (polar diagram) shaped like a figure '8'. To calculate the length of a half wave receiving dipole at a given frequency the formula:

$$\frac{468}{\text{frequency (MHz)}}$$

will give the answer in feet.

This formula gives the electrical half wavelength as applied to an aerial structure and is shorter than the physical half wavelength in free space. The reason for this shortening factor is related to the velocity of propagation within the aerial structure, the effects of the aerial elements and its supporting insulation actually causes an incident wave to slow down. A further complication arises since the diameter of the aerial element is a related factor to be included in accurate calculation for correct resonance. For our purposes however it is sufficient to regard a correction factor of 5% as related to the free space half wavelength. To find the correct free space half wavelength the following formula may be used:

$$\frac{492}{\text{f (MHz)}} \quad \text{(answer in feet)}$$

In all calculations relating to aerial construction and element lengths the '468' formula must be used but to establish correct spacing between stacked arrays the '492' formula should be used. The velocity factor of some coaxial cables can be much higher than the 5% factor as noted above and in such cases the '492' formula should be used and multiplied by the velocity factor of the specific cable eg:

$$\frac{492}{\text{f (MHz)}} \times .85 \text{ (answer in feet)}$$

This will give an electrical half wavelength of a section of coaxial cable having a velocity factor of 0.85.

The signal induced within the dipole is usually fed to the

receiver by means of coaxial cable. The connection is made at
the centre of the half wave dipole. At this point the characteristics impedance is in the order of 70–75 ohms and to ensure
the minimum loss of signal a coaxial cable also having a
characteristic impedance of 75 ohms should be used, thus
ensuring a good match and the maximum transfer of signal
energy from the dipole into the cable. If there is a mismatch
at this point a loss of signal will occur. In Europe and the
Americas, ribbon feeder is commonly used with a somewhat
higher impedance than the coaxial cable, mentioned and is
connected to a dipole assembly having an appropriately
higher centre impedance. The same general comments apply
to this feeder as to coaxial cable.

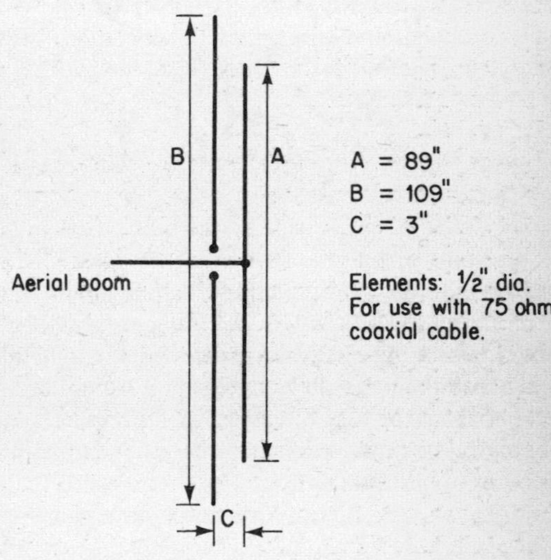

Wideband Band 1 dipole (ch. E2-4)

The half wave dipole is satisfactory when operated in a field
of high signal strength but enhanced gain is required when
signal strengths fall to a low value. An improvement in gain can
be obtained by mounting another longer element a distance
from the dipole (usually ¼ to ½ wavelength spacing). This

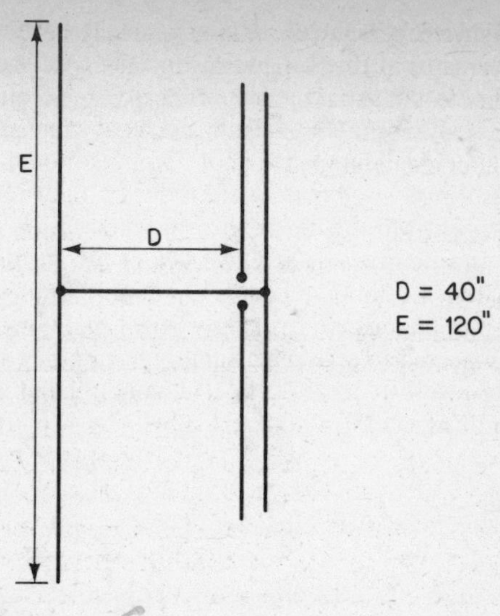

Wideband Band 1 2-element array

element reflects signals back to the dipole thus increasing the signal voltage within the dipole. Further elements, called directors, shorter than the dipole, can be mounted in front of the dipole and together with the reflector provides a considerable increase in signal gain. With the increasing number of elements so the polar diagram changes, from that of omni-directional (in the case of vertical polarisation), to one showing a forward lobe in one particular direction and with minimum responses in other directions. The sharpness of the main forward lobe will depend upon the number of elements within the aerial system. A horizontally polarised aerial will also show an enhanced forward pickup lobe when the elements are increased in number. Unfortunately when additional elements are added to an aerial system the centre impedance of the dipole will drop from the initial 75 ohms to a considerably lower figure. To maintain a good match to the cable, the dipole impedance may be again increased by folding the dipole or by various other means. A typical array, designed for one channel, operation will have a bandwidth of approximately 5 − 7 MHz, this figure depending upon the frequency in use, the number of elements in the aerial array and the type and thickness of the dipole. An aerial having a reflector and a number of directors is known as a Yagi, the dipole being the active element, the other elements passive.

For the purpose of long distance television reception when a number of channels have to be covered efficiently, a wide band aerial should be used. The design of such an aerial follows a basic formula of tuning the various elements comprising the array to certain frequencies within the operating bandwidths. Usually the directors are tuned to the high frequency end, the dipole midway, and the reflector assembly to the lower frequency end of the band. If a number of directors are used this will tend to lift the gain at the high frequency end some-what higher than that of the low frequency end. As signal losses tend to increase with frequency, the increased high frequency gain will tend to equal out such losses and give a more level response over the band.

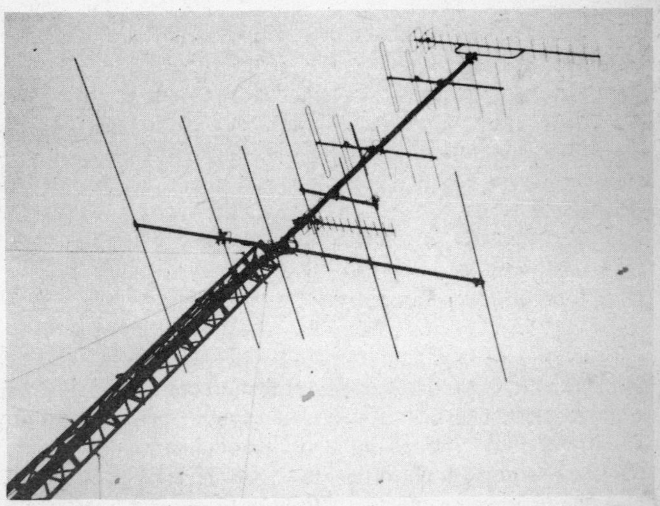

An early TV-DX installation featuring a 5-element wideband Band 1 array

Several Wide Band 1 designs are shown ranging from a simple dipole to more elaborate designs. Three of the designs feature variations of the patented Antiference 'Tru-match' system in which the active dipole has a closely coupled passive dipole but cut to the high frequency part of the required bandwidth. The driven dipole being longer than half wave is inductive at the low frequency end whilst the parasitic dipole is less than half wave and so is capacitive at the high frequency end thus producing a better match over the design bandwidth. The action of the two dipoles produces at resonance a close match to 75 ohms and off resonance the reactive swings are reduced due to the self correcting properties of each dipole element. The two element version gives a degree of forward gain over the bandwidth, the four element however gives increased gain particularly over the high frequency end of the band. A folded dipole is used for an improved feeder/aerial match since close spacing is used with the reflector and first director.

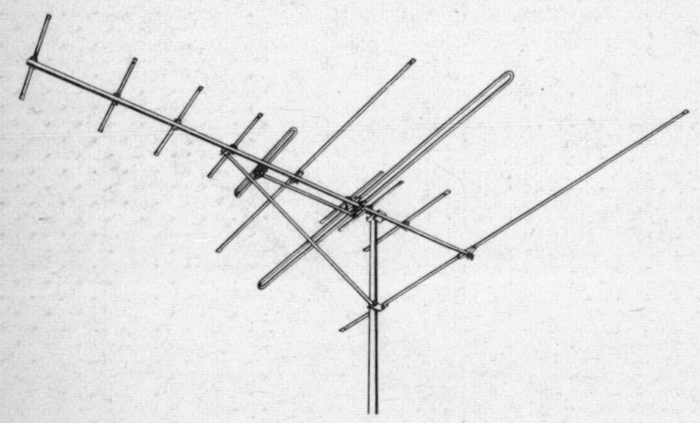

The MH308 1/3 array (Antiference Ltd)

Jaybeam, one of the UK's foremost manufacturers of television aerials suggested another design for a three element wide band system which uses a single dipole only and incorporates an increased bandwidth reaching down to 40 MHz. The range of wide band yagis is completed with the five element array based on the Jaybeam design but slightly modified for chs. E2–4 inclusive coverage. Band 2 (TV) dimensions are also given. For

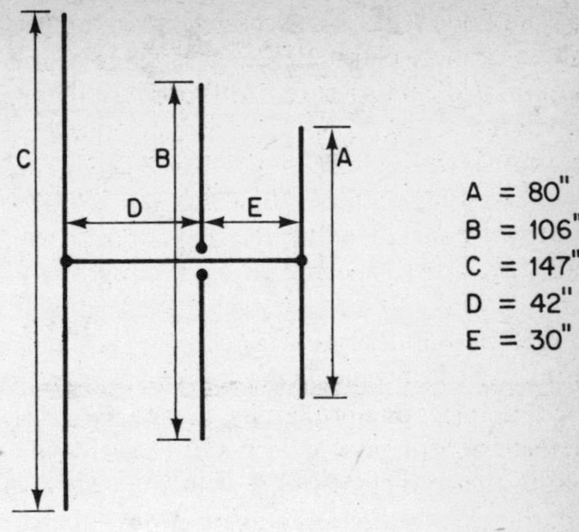

Wideband Band 1 array by Jaybeam (40-70 MHz)

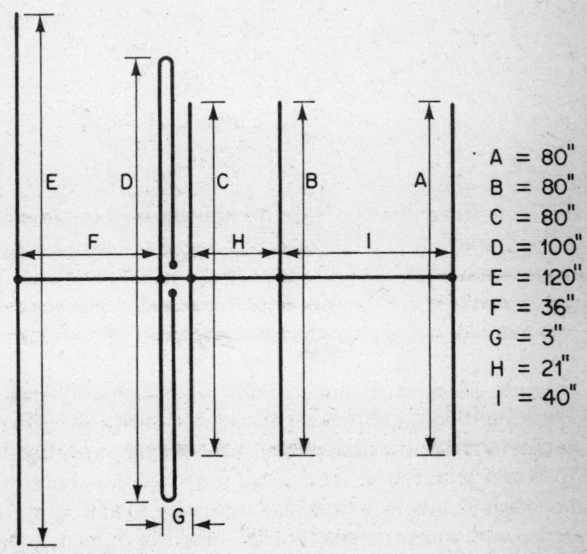

Wideband Band 1 array (47-70 MHz)

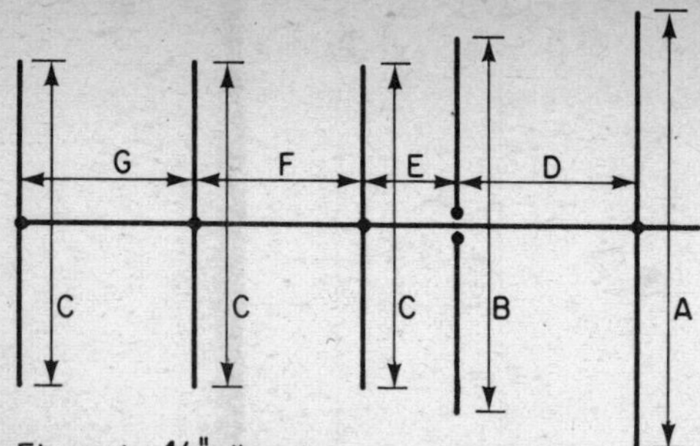

Elements: ½" diameter.
For use with 75 ohm coaxial cable.

Dimension	Length in inches	
	Band 1 45–70 MHz	Band 2 70–93 MHz
A	121	80
B	99	68
C	80	60
D	40	34
E	24	18
F	40	32
G	42	33

Wideband aerial design

use with 300 ohms ribbon feeder the dipole should be folded
giving a four times increase in impedance. At Band 3 and
higher such wide band arrays prove easier to match accurately
into the appropriate feeder and often small matching transformers (Baluns) are included.

A simple aerial which gives an omni-directional coverage is
illustrated. This consists of two dipoles, cut to ch. E3 mounted at 90° to each other, phased and matched together into a
75 ohms output. The aerial when mounted horizontally provides
a circular receiving pattern due to the combination of the 2"
figure of 8" polar diagrams of the dipoles. This aerial is
recommended for Band 1 Meteor Shower/Scatter (MS) and
Sporadic E use for the person unable to erect large arrays. This

array may be constructed using the wide band dipole (Tru-match system) with an improved match over the chs. E2–4 bandwidth, rather than the straight single dipoles as shown. The matching and phasing harness remains unaltered. A commercial wide band ferrite cored combiner could be used to couple the 2 dipoles in place of the cable harness ensuring the 2 feeders to the transformer are of equal length.

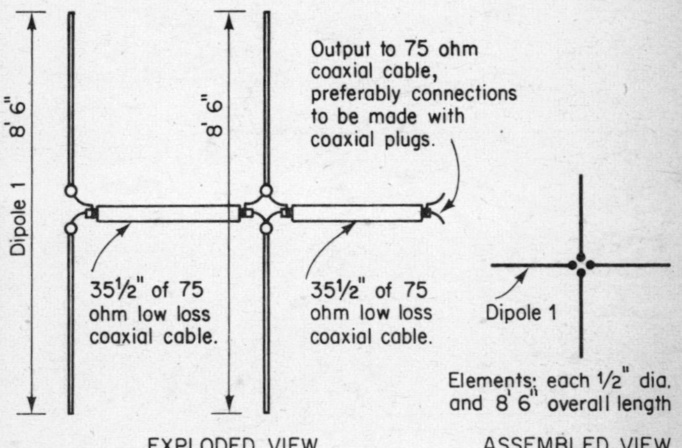

Omni-directional aerial (Band 1)

An advanced omni-directional wideband Band 1 array with reflector system (South West Aerial Systems)

Another type of array that has a superior performance for Band 1 reception is the Log Periodic due to a level gain throughout the design bandwidth, excellent matching, low VSWR and a

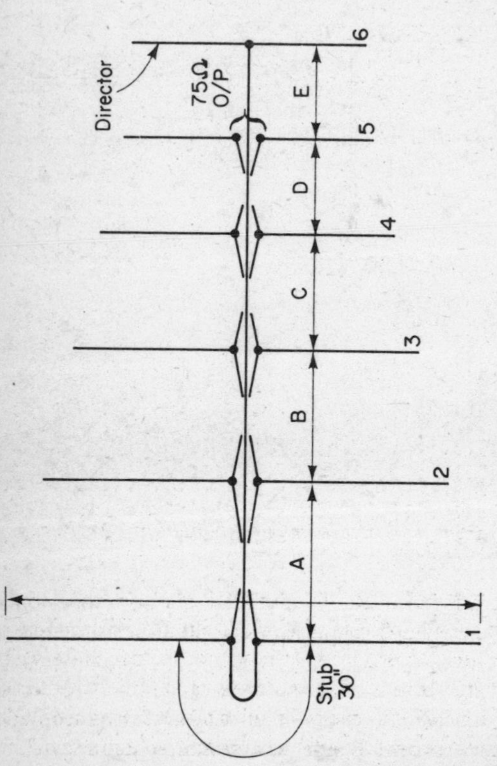

Log Periodic array for chs. E2-4 inclusive

very clean polar diagram free from minor lobes. The 6 element as shown features a director which can be omitted to reduce the overall boom length. Connections between each dipole can comprise copper cable and soldered to tags on each dipole mounting. The tags and the immediate area on the dipole element must be greased (ideally silicone) in the interests of minimising electrolytic corrosion. The 75 ohm coaxial feeder should be either taped underneath the boom and for the total length to the rear or passed inside the actual boom element itself. Care should be taken to avoid the boom filling with water. This type of array is not the easiest to manufacture by the amateur enthusiast and the beginner into the hobby would be advised to initially concentrate on the simpler Yagi systems detailed earlier.

A wideband Band 1/3 colinear model MH473 (Antiference Ltd)

Wide band arrays for Band 3 and UHF are now commonly available, having high gain and an excellent performance generally. The field is so wide that the would-be purchaser is advised to contact the various manufacturers or suppliers for imformation. Within the UK several manufacturers produce Wide Band 3 arrays, such as the J Beam 'Astrabeam', which is available in versions up to a double 11, possessing a high gain and a sharp forward pickup lobe.

With increasing activity at UHF competition between manufacturers has heralded greatly improved arrays departing somewhat from the conventional Yagi appearance with director assemblies following Continental European practice.

Several manufacturers currently produce versions of the multiple director unit array and although the appearance may differ somewhat the basic theory remains unaltered. The multiple director unit is an attempt to obtain the performance of a stacked array by supporting the director elements in a position that would be occupied by the individual directors of the separate stacked aerials. The advantage with the system is that a very high gain may be obtained using only one aerial boom and without the need for restricting cable harness, a gain that would be unobtainable from a conventional Yagi with straight director elements. Illustrations show the types of multiple director array now available in the United Kingdom.

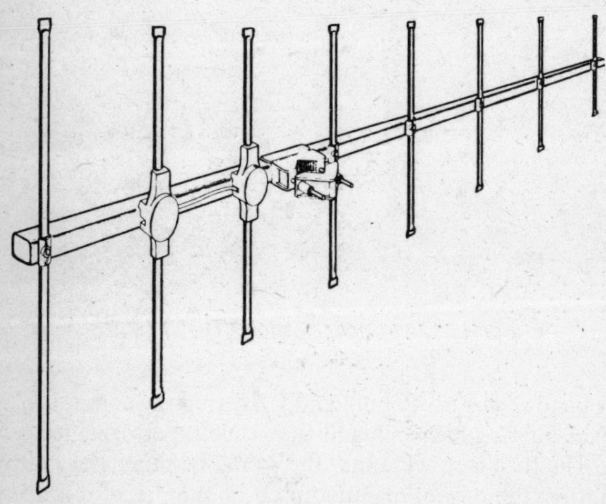

A wideband 3 array featuring twin driven dipoles (Model AC8, Wolsey Electronics)

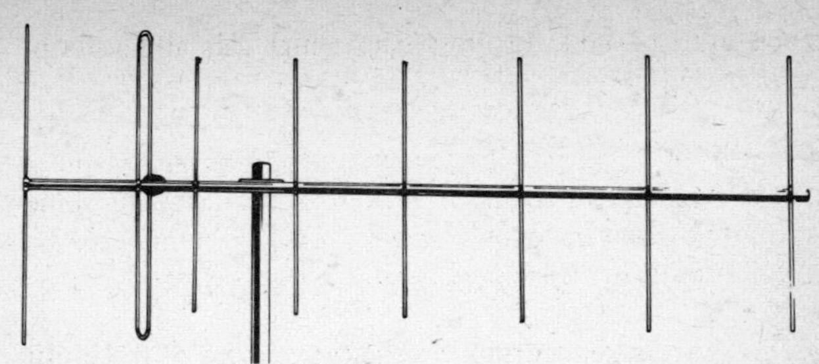

A wideband 3 8-element array with single dipole and internal balun (Jaybeam Ltd)

A standard UHF 18-element wide spaced Yagi
(SG18, Aerialite Ltd)

A Group C standard in-line Yagi (QR18, Wolsey Electronics)

The PBM18, an 18-element array featuring a modified slot dipole structure (Jaybeam Ltd)

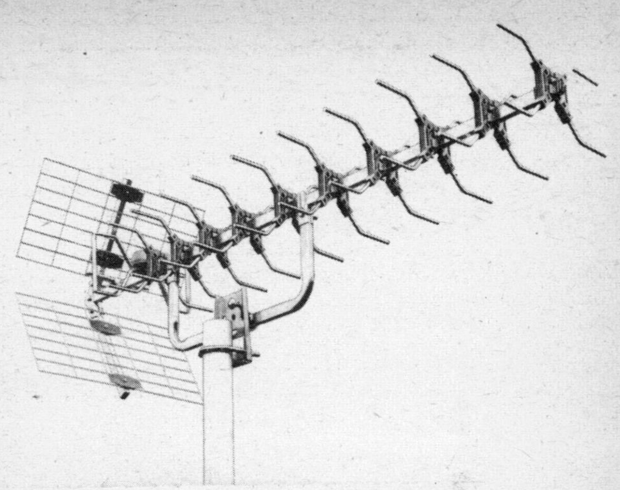

A multiple director array, the JBX10 (Jaybeam Ltd)

*The HG36 high gain multiple director system
(Wolsey Electronics)*

Available in both grouped and wideband, the XG21 multiple director high gain UHF array used full wave director units (Antiference Ltd)

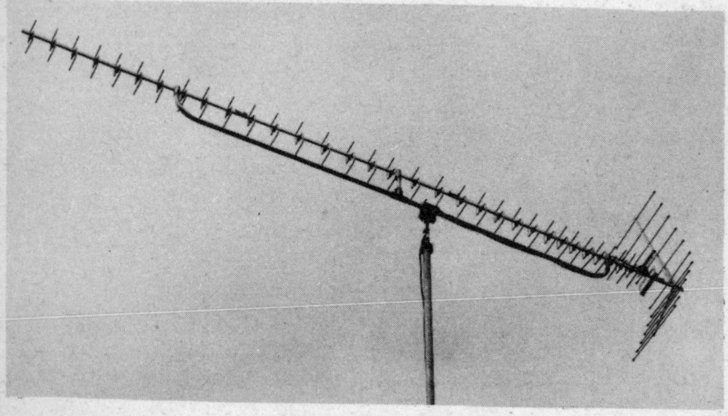

A multiple director UHF array using ½ wave elements (GSP98, Aerialite Ltd)

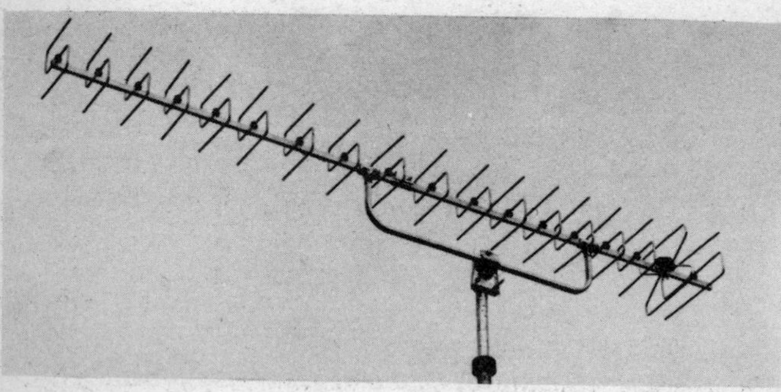

A lower gain system with ½ wave elements (GSP50, Aerialite Ltd)

The Vorta wideband UHF array (premier Industries [Cheltenham] Ltd)

A West German wideband UHF array, the Fuba XC391 (imported by Audio Workshops)

The UHF transmitter coverage has been arranged in the United Kingdom so that in any one service area the transmissions originate from one transmitter site and that these channels are restricted into a channel grouping — this is in the interests of minimising interference with adjacent service areas and for the optimum reception quality on the home receiver. The channel grouping currently employed is:

Group A	ch. 21—34
Group B	ch. 39—53
Group C/D	ch. 48—68
Group E	ch. 39—68

Continental European designs tend to be divided into a basic Band 4 — ch. 21 — 37; Band 5 — ch. 39 — 68, and further sub-divisions. One interesting feature is that a wide band UHF multiple director design is usually provided covering the whole of Bands 4 and 5. British arrays of the multiple director range have tended to be restricted into the various channel groupings as detailed above but at the time of writing several manufacturers have now introduced wide band versions.

The wide band UHF aerial situation remained static for some years in the UK with generally the Log Periodic system being the only array with coverage in both bands. This array differs from the Yagi in that all elements are active and although the gain is somewhat lower than an equivalent Yagi with the same number of elements, it produces an extremely clean polar diagram with very low side lobe radiation patterns. The bandwidth and gain figures maintain a relatively level response throughout its designed frequency limits. The first 'breakthrough' in higher gain wide band UHF systems was the introduction of a stacked bowtie array, an aerial that had been very popular in the United States for many years. The array (marketed by Wolsey) has a quoted gain ranging between 13.2 to 15.5 dB compared with the typical Log Periodic at a level 9dB or so. When equipped with a matching wide band UHF amplifier the long distance enthusiast had the best compromise for his activities. Antiference have also introduced wide banded versions for the whole UHF band in their 'XG' range and possessing a

A wideband UHF Log Periodic aerial type LBM2 (Jaybeam Ltd)

very high gain, characteristic of the multiple director system, and several imported '✕' director aerials are now available on the UK market.

An entrant to the wideband UHF field that unfortunately enjoyed a brief commercial life was the Short Backfire. This aerial was based upon a Czechoslovakian design (although there are claims that the design originated in the U.S.A.) and departs from the conventional UHF array in common use. The array featured a good bandwidth performance particularly when a 7 element director extension was fitted and a stacked array exhibited phenomenally deep nulls in its polar diagram allowing unwanted signals to be 'lost' and with 'on beam' gains of 15dB in Group A to 19dB at the high frequency end of Group C/D. A large reflector screen some 24 inches square in conjunction with a smaller reflector form a cavity resonator, signal pickup is with the shallow '✕' dipole between the two reflectors. The gain rises from a low of 12dB at the low frequency end to above 16dB at the high frequency end of the UHF band. An aerial that is used extensively for deep fringe working in the United States is the Dish Reflector. A high gain array it usually features a single or stacked bowtie facing in the dish, the dish itself often measuring some 7 foot high and 5 foot wide.

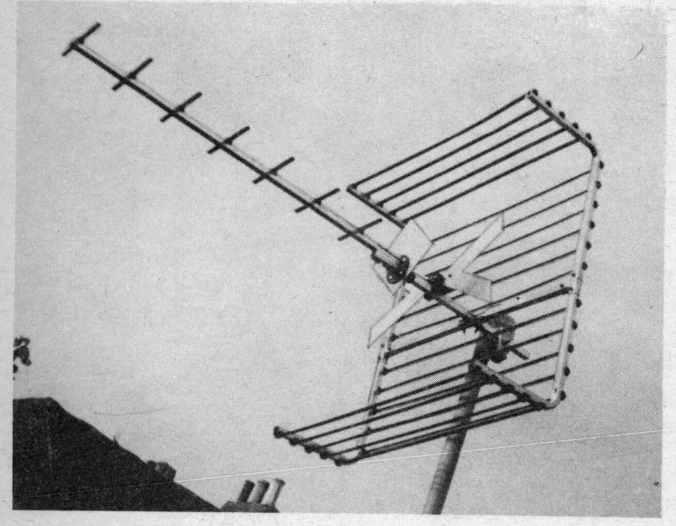

An unusual wideband UHF array, the Short Backfire
(formally Telerection Ltd)

The gain of an aerial can be increased by combining its signal output with and other similar aerial mounted in close proximity. When such aerials are combined (or stacked) an alteration also occurs to the polar diagram. For vertically polarised arrays stacked side-by-side the forward pickup lobe is considerably sharpened. Horizontal arrays may be stacked side-by-side or above each other. Sideways stacking sharpens the forward pickup lobe, but stacking vertically above each other only marginally narrows the forward pickup lobe, but it does reduce the vertical acceptance angle. As signals can at times be received via Meteor Scatter in Band 3, sideways stacking is to be preferred, to exploit the wider vertical acceptance angle.

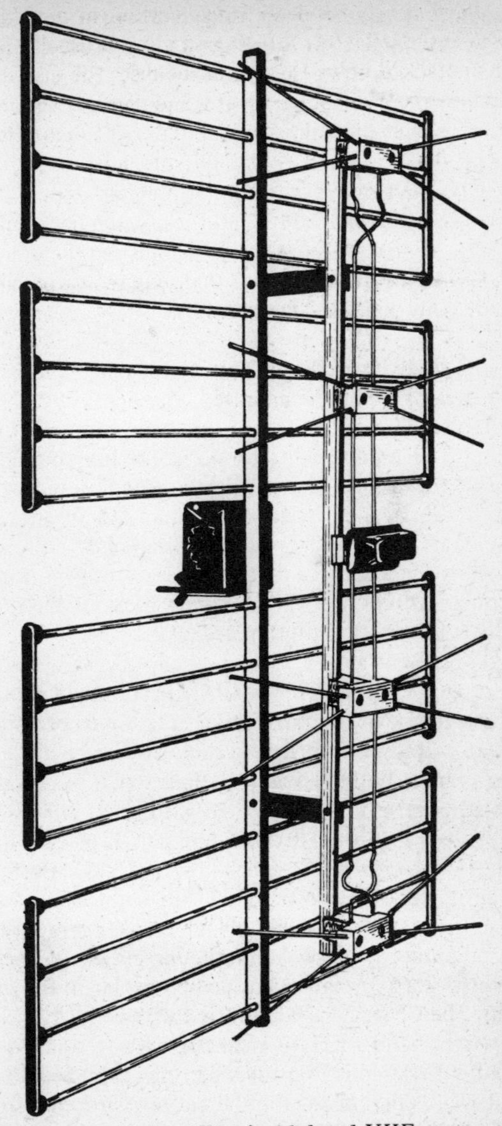

The Wolsey 'Colour King' wideband UHF array using stacked dipole assemblies (Wolsey Electronics)

Unfortunately there is at present no stacking system or harness
that allows stacking of two similar wide band UHF arrays and
still maintains the original design bandwidth. There are
available, however, wideband ferrite cored aerial combiners for
VHF/UHF use but these unfortunately exhibit slight insertion
losses, typically 0.5dB and 1dB VHF and UHF respectively.
Group UHF arrays can be stacked to obtain an extra 3dB
power gain over the bandwidth and the spacing between the
arrays can provide a useful method of adjusting the polar
diagram to give nulls in directions that may be introducing
interference into the receiving system.

As a general receiving standard, for long distance television
reception, aerials should be mounted horizontally. With the
exception of the United Kingdom and some countries in
Europe, most high powered transmitters use horizontal polarisation. Due to the changing polarisation often noted with
Sporadic E signals, an array can be mounted to advantage
vertically in areas where strong local horizontally polarised
transmissions occur, or vice versa if the local transmitter uses
vertical polarisation. This will result in a considerable reduction of local transmitter interference.

The use of circular polarisation rather than a single plane
polarisation in television transmission is likely to become
widespread through North America and elsewhere during the
next few years at both VHF and UHF. Circular polarisation
(CP) has advantages in built up areas with ghost reduction and
with improving the results from indoor aerials of random polarisation. Fringe area performance however may deteriorate
if the transmitter power or its aerial gain is not increased. CP
signals are compatible with any plane polarised array, the
array responding equally well mounted in either the vertical
or horizontal plane provided it is perpendicular to the propagation path. The CP signal is transmitted with a clockwise rotating field (right hand —RH) or an anti-clockwise rotating field
(left hand — LH) — this rotation when viewed from the aerial
looking towards the propagation direction. Optimum signal
output from a CP array occurs when both transmitting and
receiving arrays are of the same rotational sense, minimal

An experimental Helical Group A UHF array

performance is obtained when using CP arrays of opposite rotational sense.

Gaining popularity are 'active aerials'. These are basically a compressed array featuring inductive loading within specially designed containers resulting in reduced dimensions for Band 1 and 2 arrays. Dimensions of these containers are such that Band 3 and UHF arrays can be made to their normal resonant lengths. Typical sizes for these arrays are in the order of 80–100 cms. for VHF types and reducing to perhaps 40cms. for UHF only systems, and appearance is usually that of a flat horizontal disc or fluted dome. By their very name 'active aerials' contain integral amplification stages to give both matching and gain, versions without integral amplifiers have outputs ranging to –6dB in Band 1 and decreasing to --2dB at UHF. Typical outputs from such compressed 'active aerials' are +15-20dB at VHF, +25-30dB at UHF, related to a ½ wave dipole.

A wideband Band 1 Log Periodic on a roof in Holland

A wideband VHF/UHF 'active aerial', intended for marine use. Some 20" in diameter, it features an integral wideband amplifier of some 27 dB gain, noise 5.5 dB and coverage of 40-860 MHz (Aeranamics Electronics Ltd, Peterhead)

Where possible the aerials should be mounted in the open and away from obstructions. The type and height of a mast supporting an aerial will depend largely upon local conditions, hills, trees and neighbouring terrain, the attitude of planning authorities and personal economics. Sporadic E signals can be

An elaborate TV-DX receiving system.
Note the satellite dish on the mast

well received on low aerials although it is preferable to lift
such an array to a minimum of 25 feet. Tropospheric reception
necessitates a greater degree of aerial selection and positioning
with the use of high gain arrays at the maximum height possible.
For both Sporadic E and Tropospheric reception, a slight up-
ward tilt of 5–10º to the receiving aerial is often beneficial.

An impressive TV-DX installation in the Netherlands

For multi element arrays, arrangements should be made to
rotate the system either by mechanical or electrical means;
rotor units are now available from various aerial suppliers.
Precautions should be taken to allow sufficient cable slackness
at points between moving and fixed parts of the mast, in order
to avoid cable stress and damage.

The maximum space possible should be allowed between
separate aerials on the same mast in the interests of avoiding

The author's former aerial mast.
Note the UHF dish array imported from the USA

signal absorption and other undesirable effects. When mounting aerials upon a mast care should be taken to observe wind loading and other stresses that may occur should too many large arrays be fitted.

Once the aerial has extracted a weak signal from the ether it passes it to the receiver through the feeder cable. This will be of two types: coaxial cable or ribbon feeder. It is essential that the feeder has very low loss figures. Coaxial cable has an inner conductor, which with low loss cable is a single solid conductor, the surrounding dielectric consisting of cellular polythene or airspaced ribbon polythene. Ribbon feeder varies somewhat, often with a hollow tube construction, at times foam filled. Greater care has to be taken when fitting ribbon feeder to a mast, it being held away from the actual metalwork to avoid losses. Certain types of ribbon feeder have a very low signal loss when dry, unfortunately the losses tend to rise when the

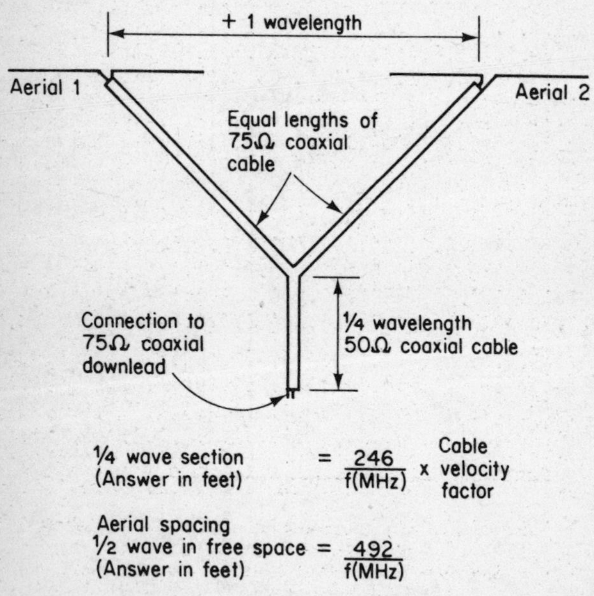

Narrow band aerial stacking using cable harness

Typical Cable Characteristics

| | | Inner conductor | | Polythene dielectric | Cable diam (mm) | Capacitance (pf/metre) | Nominal attenuation dB/100 metres | | | | |
	Type	No. of wires	Wire diam (mm)				50 MHz	100 MHz	200 MHz	650 MHz	850 MHz
AERIALITE COAXIAL CABLE	4202	7	0.25	Cellular	5.10	56	7.2	10.2	14.5	17.2	31.4
	4203	1	1.12	Cellular	7.25	56	5.2	7.4	10.4	19.6	27.8
	4205	1	1.25	Aeraxial	7.15	57	4.6	6.5	9.2	17.2	19.9
	4209	1	1.40	Aeraxial	8.00	56	4.0	5.7	8.0	14.8	17.0
STOLLE 300 OHM RIBBON CABLE	2010	2x7	0.28	PE				4.5	7.0	14.0 @ 600 MHz	17.0 @ 800 MHz

ribbon becomes wet. The signal loss within a feeder system will also rise with an increase in frequency. Whichever type of feeder is used, wise policy will dictate that the best quality is fitted. The table shows characteristics of four different types of coaxial cable often used for television aerial feeder. The stranded inner conductor coaxial cable has by far the greatest signal loss and should be avoided at all frequencies when used in weak signal work.

AERIAL TERMINOLOGY

Gain

This denotes the efficiency of a receiving array in terms of an improvement in signal output as compared (usually) with a half wave dipole. The power gain of an aerial is measured in decibels (dB). An improvement in gain of 3dB denotes a doubling in signal output and usually the actual physical dimension of a receiving array is doubled to obtain the 3dB gain.

Decibel

A logarithmic unit of measurement denoting signal levels, gain and loss. Decibel ratio charts are given in many radio reference books.

Signal voltage

This is normally expressed in microvolts or millivolts (μV and mV). One mV = 1000 μV. Often expressed in dBmV or dBμV: eg 0dBmV = 1mV.

Front to Back Ratio

This ratio is measured in dB and indicates the efficiency in receiving the required forward approaching signal compared with the efficiency in rejecting the unwanted signal arriving over the rear (reflector) end of the aerial.

Beamwidth/Acceptance Angle

This is an indication of the directional ability of a receiving aerial. The beamwidth is usually measured between the half power (3dB) points on the required forward approaching signal.

Voltage Standing Wave Ratio (VSWR)

The ratio of the maximum voltage set up along a line and compared to the minimum voltage and is an index of the matching/mismatching properties of a line, the aerial and terminating receiver. A well designed commercial array will often present a VSWR of 1.5/1 or better whereas a home constructed Band 1 may display a VSWR of 2/1 or worse, depending on the efficiency of design and bandwidth covered.

Polar Diagram

The polar diagram of an array is the response pattern as the array is rotated through 360° whilst receiving a signal from a specific direction. If the signal output from the array is plotted on a 360° graph, its measurements corresponding to the aerial position a drawing is obtained showing the characteristics of the array such as beamwidth, front/back and the side lobe pattern.

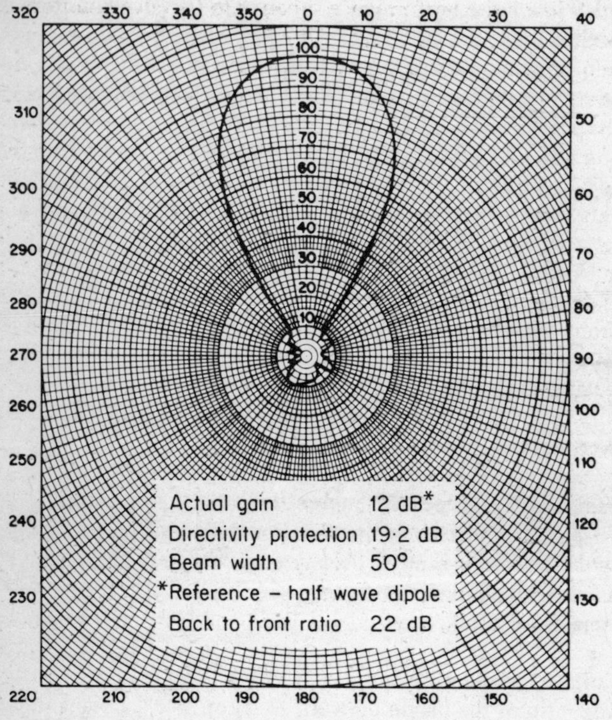

Model SG10 standard channel group UHF TV aerial
Wide–spaced 10 element Yagi
Polar diagram 'B' group E–plane (mid–channel)

Aerialite technical data sheet

AERIAL AMPLIFIERS

The signal strengths encountered with very weak distant
television transmitters are such that they often lie below that of
the receiver's threshold sensitivity and are unusable. At such
times the use of an aerial amplifier is invaluable to lift the weak
signal to a level that the receiver can accept and produce a
recognisable image.

The transistor aerial amplifier has now come into universal use
with a low noise performance superior to its valve counterpart.
The low input impedance allows good matching (and a conse-
quent minimum noise figure) from a 75 ohm aerial feeder, and
assists with the design of amplifiers requiring a wide bandwidth.
With the use of wide band aerials (which have a gain inversely
proportional to bandwidth), an aerial amplifier of similar coverage allows for an efficient receiving system. As a general rule
noise tends to increase with frequency for both valve and
transistor circuits.

To maintain the lowest noise performance possible on a wide
band amplifier, which will tend to have a higher noise for a given
transistor than that of a narrow band amplifier, the use of
UHF devices at both UHF and VHF is advised.

Several circuit designs have been given using the BF 180 in use
at VHF and UHF with gains in a wide band circuit of 16dB at
Band 1,14dB at Band 3, 12dB at Band 4 falling to 10dB at
Band 5. The expected noise figures will vary from 2dB at
Band 1 to 6dB at Band 5. As a general rule stages may be
cascaded to give increased gain but care must be taken in the
interests of stability and a level gain/bandwidth performance.

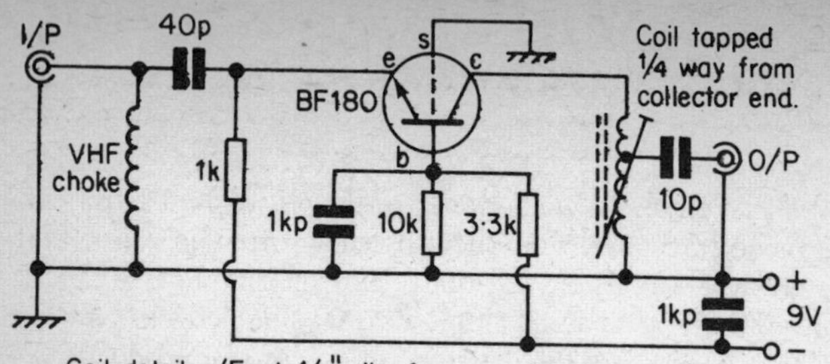

Coil details: (Each 1/4" dia. formers with dust core)
ChE2: 12 turns 26g enam close spaced.
ChE3: 10   "   "   "   "   "
ChE4: 8   "   "   "   "   "
Band 2 (TV): 6 turns 26g enam close spaced.
VHF Choke: 1/8" dia. former 12 turns 28g enam close sp.

AERIAL AMPLIFIER VHF.

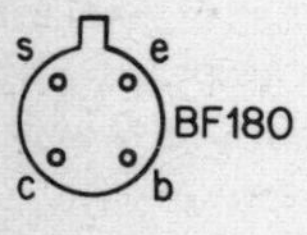

Coil details:
L1: tuned to ChE4.
L3: tuned to ChE2.
L2: 3 turns 26g PVC
    wound over
    'dead end' of L1.

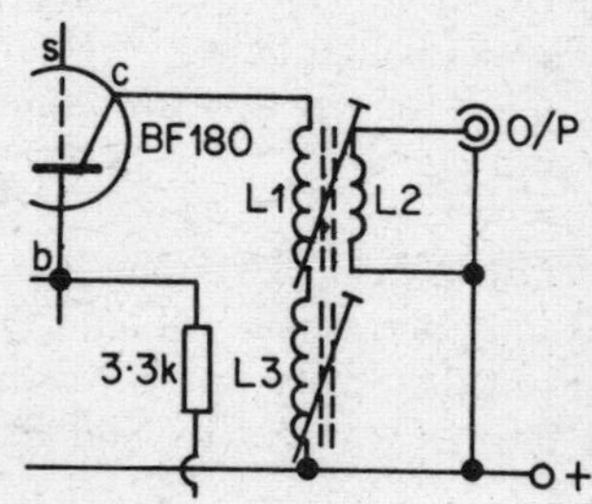

ALTERNATIVE WIDE BAND 1 CIRCUIT.

Coil details:
L1: 1/4" dia former, 5
    turns 20g enam
    spaced over 1/2".
L2: 2 turns 26g PVC
    wound between
    turns of L1 starting
    1 1/2 turns from
    collector end.

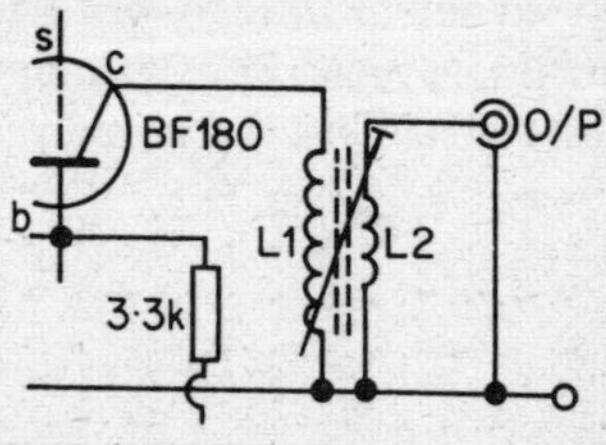

WIDE BAND 3 CIRCUIT (170–210 MHz).

Aerial amplifier VHF

CT: 4–10pf low loss trimmer.
Tuned Line: 0·048" dia. copper.
Input Coupling to emitter: 5pf.

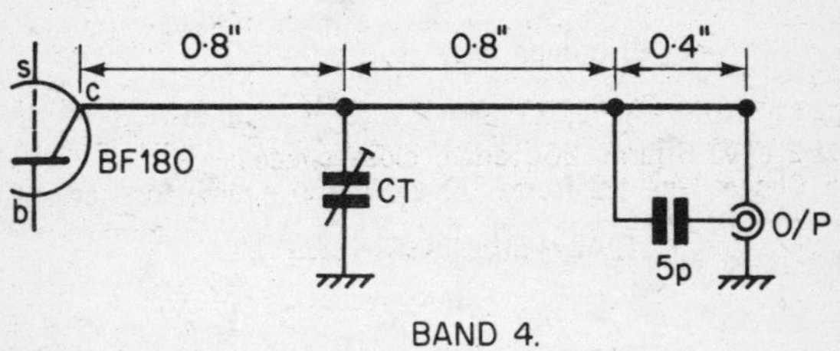

BAND 4.

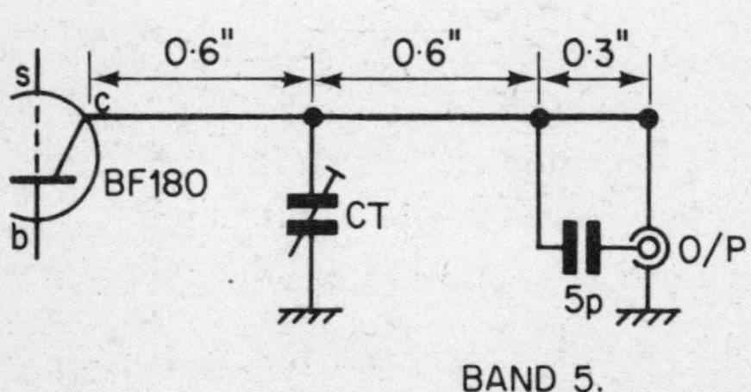

BAND 5.

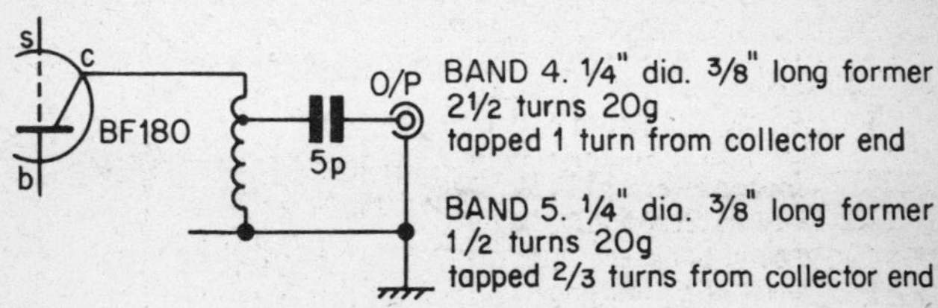

BAND 4. ¼" dia. ⅜" long former
2½ turns 20g
tapped 1 turn from collector end

BAND 5. ¼" dia. ⅜" long former
1/2 turns 20g
tapped ⅔ turns from collector end

ALTERNATIVE COIL FOR UHF

Aerial amplifier UHF

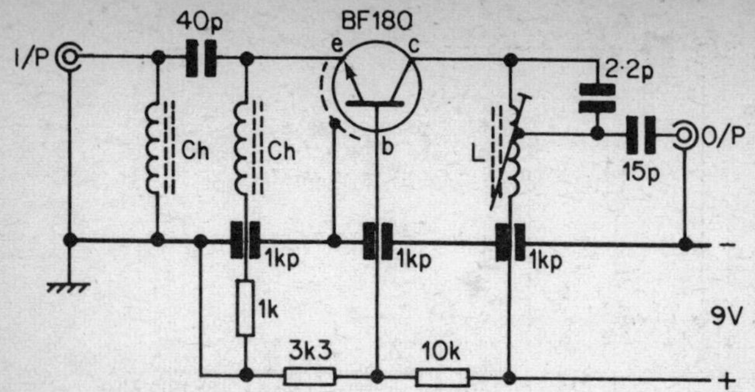

L: 10 turns close spaced 26g tapped at 5 turns ¼" coil former.

Ch: Radio Spares VHF choke.

Band 1 aerial amplifier

Alternative circuits are given for later production transistor types BF272, BF362 and a skeleton circuit for the BF479. This latter circuit may be used at either VHF or UHF in wide or narrow band mode. The gain to be expected in Band 1 for a single stage will approach 20dB and with a noise figure of under 3dB.

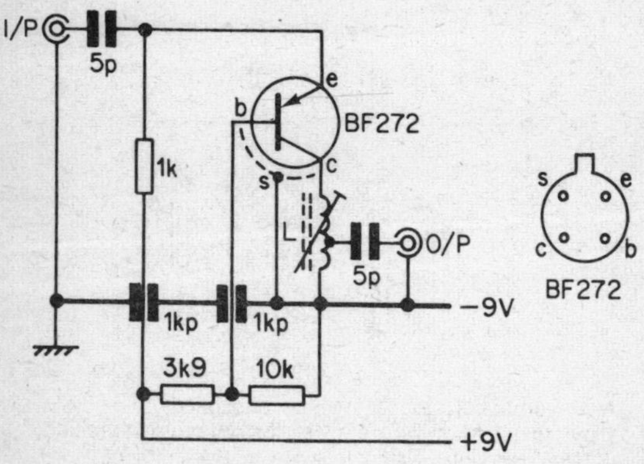

An alternative transistor for use in VHF/UHF amplifiers

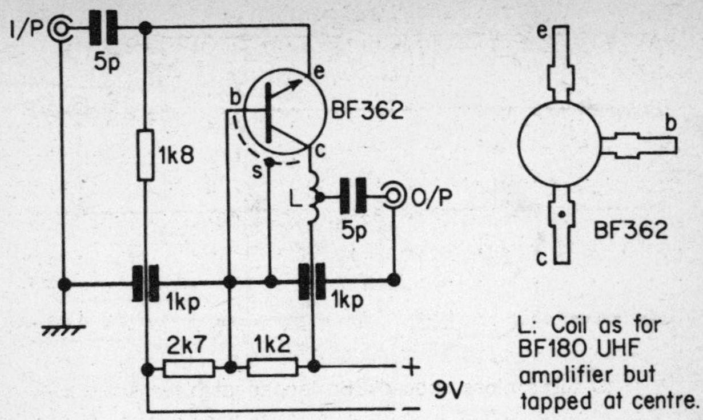

UHF aerial amplifier

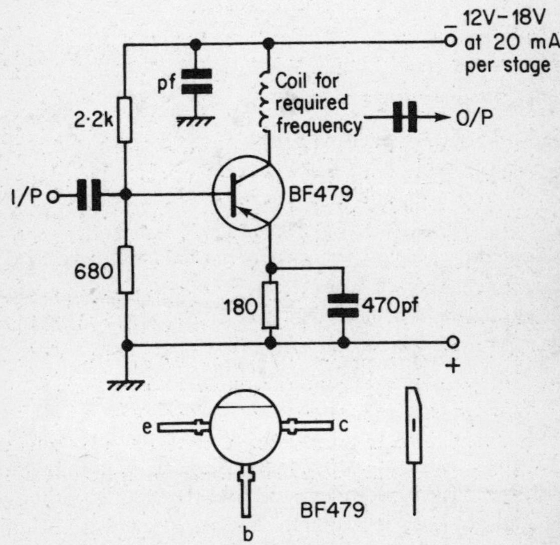

NB. Additional stages may be cascaded.
Input to 1st stage via emitter for optimum matching,
470pf decoupler fitted to base in this instance.

Skeleton preamplifier stage using BF479

A very successful low noise UHF amplifier can be made utilising the SGS BF679 and a single stage amplifier operating in a narrow band mode can realise 24dB gain with a noise figure of under 2dB.

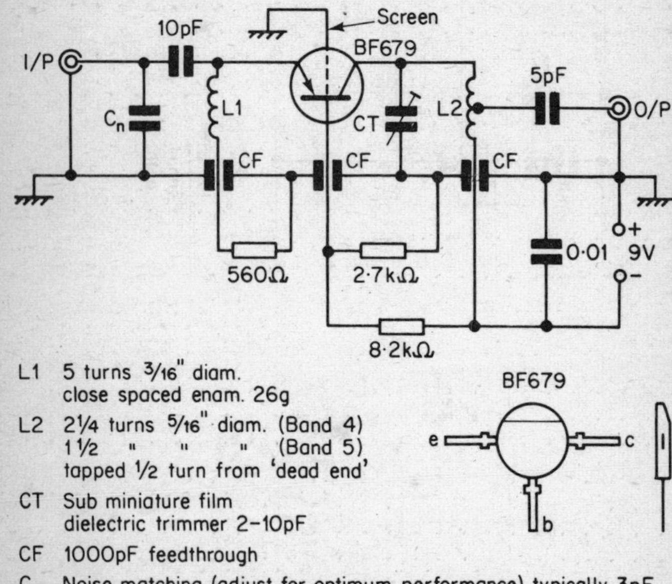

L1   5 turns ³⁄₁₆" diam.
        close spaced enam. 26g
L2   2¼ turns ⁵⁄₁₆" diam. (Band 4)
        1½   "   "   "   (Band 5)
        tapped ½ turn from 'dead end'
CT   Sub miniature film
        dielectric trimmer 2–10pF
CF   1000pF feedthrough
C_n Noise matching (adjust for optimum performance) typically 3pF
All capacitors silver mica types
All resistors low noise metal oxide ¼W at 5%

Low noise UHF amplifier

An interesting circuit that can be tuned throughout Band 1 uses 2 Field Effect transistors type 2N3823E. Tuning is carried out with the 2 varicap diodes type BA110 although other diodes could be used if available. The advantage with using an FET circuit is the freedom from cross modulation and inter modulation which can easily occur with bipolar transistors if overloading due to excessive signal input takes place. The gain of the twin stage FET aerial amplifier should be in excess of 20dB and with a very low noise figure.

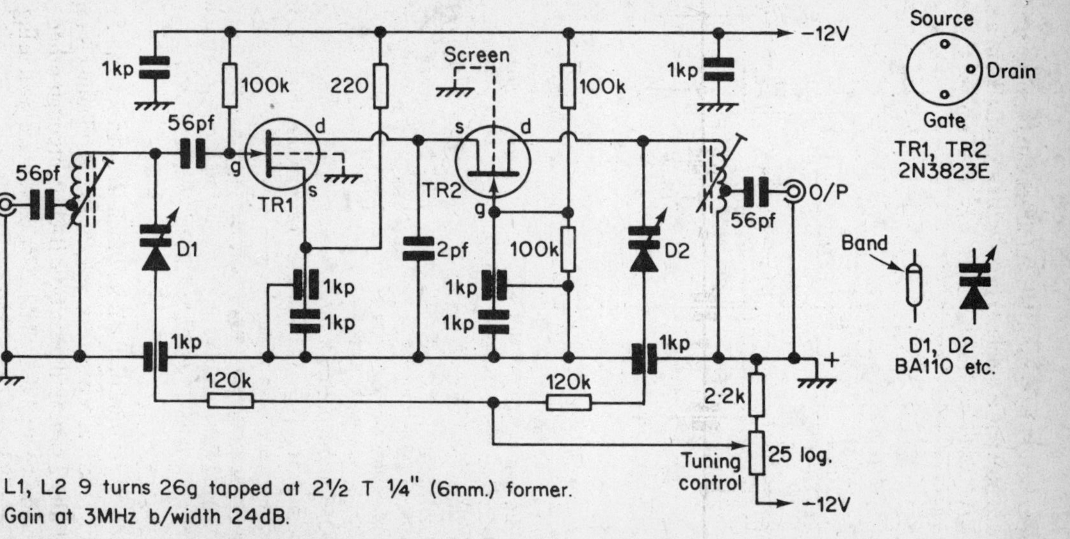

Varicap tuned FET Band 1 aerial amplifier

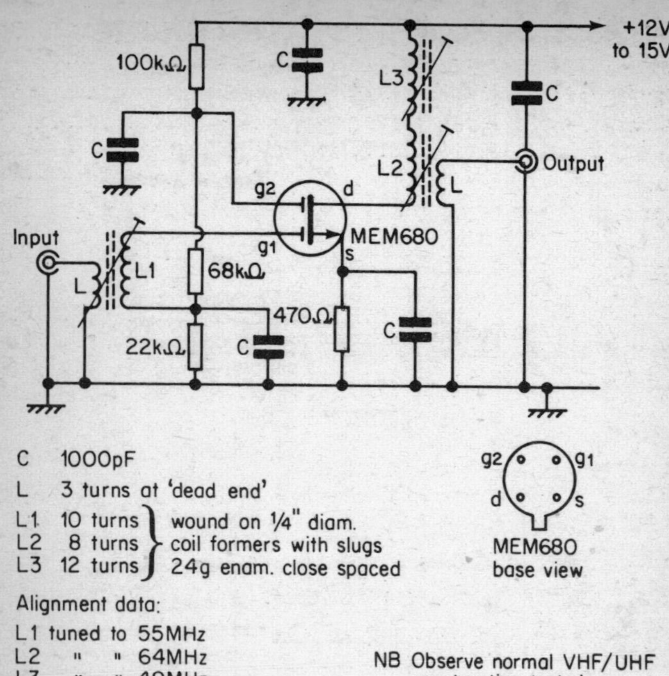

Wideband Band 1 dual gate n-channel FET amplifier

There has been a tendency in recent years for manufacturers to market wide band amplifiers covering the total spectrum used by current television transmissions namely 40-860MHz. Such amplifiers are now available having power gains of 30dB with noise figures of only 3.5dB. Mullard Ltd. have introduced a range of hybrid VHF/UHF amplifiers having various gains and noise figures, an example being the OM335 featuring a gain of 25dB with a noise figure of 5.5dB. The amplifier is encapsulated in a ceramic type coating, measuring 1 inch by ½ inch and contains all components, merely requiring 24 volts at 35mA. An alternative series of hybrid amplifiers are the SGS/ATES SH221/SH225 units. These are multiple stage amplifiers operating from a 24 volt supply at up to 35mA

and featuring gains of 18/26dB respectively, noise figures
under 5dB and a relatively level response from 30-900MHz.

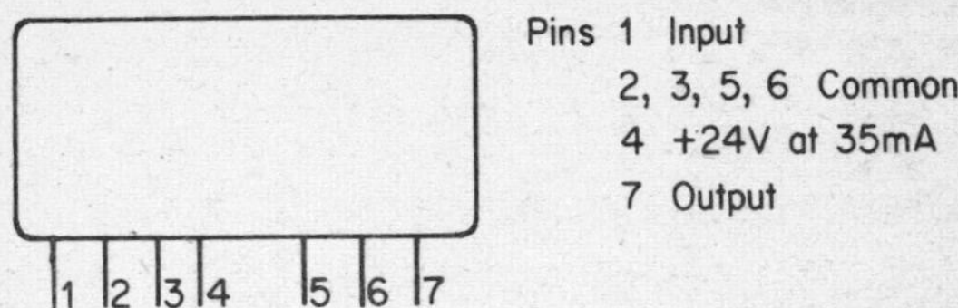

*Lead connections for the Mullard/Radiospares OM335
Hybrid amplifier*

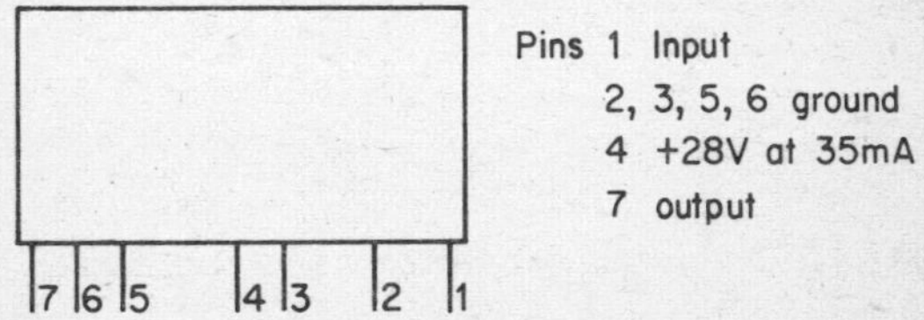

*Lead connections for the SGS/ATES Hybrid amplifiers
type SH221/SH225*

Several enthusiasts have succeeded in modifying commercial
TV tuners into high performance high gain narrow band
amplifiers. Valve tuners have been modified simply by removal
of +HT voltage to the oscillator stage and RF output being
taken from the anode of the mixer stage. Transistor UHF
tuners have similarly been modified with the removal of the
appropriate oscillator capacitor and coupling thereto, output
being taken from the mixer collector. It has been possible to
modify a varicap tuner in this way and fit the unit to the
aerial itself, tuning being carried out adjacent to the receiver.

Certain amplifiers are commercially available intended for
aerial head mounting, thus amplifying the weak signal before
any attenuation occurs in the feeder. The power is fed to the
amplifier along the common feeder line. These are available in
narrow or wideband units covering the total 40-900MHz

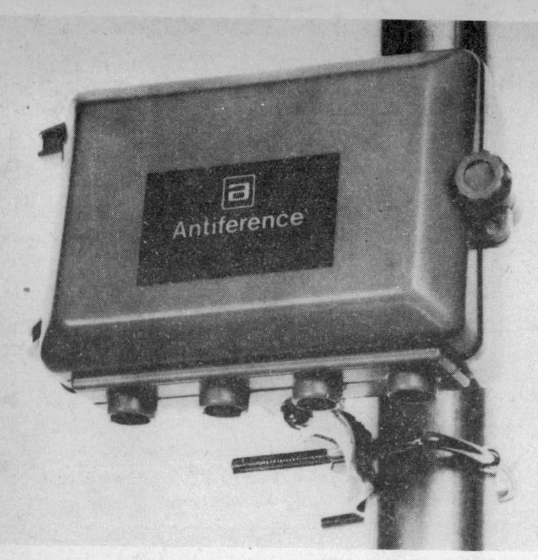

A typical masthead amplifier in its watertight case

A single stage 'group' masthead amplifier (Labgear Ltd)

spectrum, the total UHF band of 470–900MHz, each transmission group at UHF namely Groups A, B, C/D, total VHF of 40–250MHz or the separate VHF bands. Typical gains range from 12dB for a single stage unit in Group C/D, 30dB for a 3 stage wide band UHF to 22dB for a wideband 40–900MHz 2 stage unit. It is advised to contact the various manufacturers for current details of ranges available and the technical specification.

A problem sometimes encountered with high gain transistor circuits of this nature, and indeed with transistor aerial amplifiers in general is cross modulation and overloading tending to exaggerate and extend interference from very strong signal sources onto adjacent frequencies.

Amplifier gain is always quoted in decibels (dBs) and often relates to a reference level which in TV distribution systems is 1mV. Previously the dB gain discussion related to power gains and losses associated with aerials but with amplifier voltage gain the dB ratings differ somewhat. If an amplifier doubles a voltage then it has a gain of 6dB, a gain of 4 times is 12dB, and 18dB gives 8 times amplification. Assuming we are relating these gains to the 1mV reference the figure 6dBmV = 2mV (2 times 1mV), 12dBmV = 4mV, 18dBmV = 8mV etc. The noise figure indicates the extra noise (snow on vision, hiss on sound) that the amplifier itself adds when amplifying the signal. Obviously we need to amplify the weak signal but add the minimum amount of noise. Cross modulation presents itself as buzzing, excessive contrast and the presence of other local channels onto an empty channel or by the appearance of several local signals on a single local channel. The cause is an excessive signal input into the amplifier stage to a degree that the transistor(s) are unable to present a linear characteristic. Most manufacturers will quote typical handling capacities which lowers if there are several local channels. One particular manufacturer quotes a wideband UHF amplifier as handling a single signal at 44.5dBmV (167mV) but with 4 local signals the handling drops to 39.5dBmV (94mV), the unit in question being a 2 stage amplifier. A single stage single group amplifier is quoted as 28dBmV (25mV) for a single signal falling to

23 dBmV (14mV) in the presence of 4 signals. For distant signal work particularly on channels adjacent to strong local signals the importance of a low noise performance coupled with high signal handling capabilities will be obvious.

The advent of the Field Effect Transistor (FET) for VHF and UHF circuits should negate this problem — combining the advantages of valve operation with that of the bi-polar transistor. A narrow band RCA MOSFET preamplifier is featured using the 40673 device. The amplifier has an extremely low noise figure and high gain characteristics. In operation the amplifier has given an extremely good account of itself at locations close to high powered transmitters and on closely adjacent frequencies where conventional bi-polar devices would prove liable to excessive overload and cross modulation effects. For reasons of clarity protection diodes within the 40673 itself have been omitted from the circuit diagram.

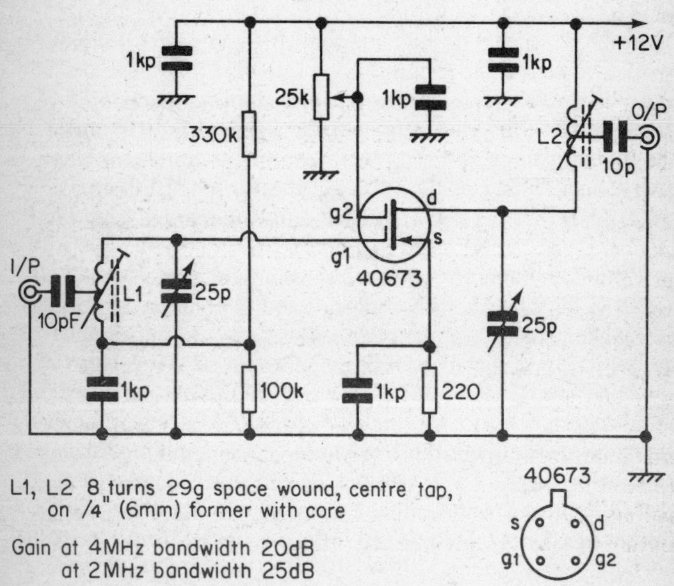

Narrow band Band 1 MOSFET aerial amplifier

PHOTOGRAPHY

A further interesting activity associated with long distance
television is the photographing of the actual received image.
Photography allows one to maintain a record of reception, com-
parisons of improvements over a period of time and facilitates
identification of unknown signals. However greater skill is
called for when photographing from the television screen but
with some practice and experience excellent results can be
obtained.

The use of a 35mm camera is recommended, initially a cheap
one should be obtained if venturing into this field for the
first time. Such a camera is less expensive from an operating
point of view, being able to take up to 40 full frame shots. It
should be able to focus down to at least three feet and have
either a shutter speed of under 1/25th or be capable of
manual operation. The reason for the 1/25th speed is that
each television picture is comprised of two interlaced fields,
there being twenty-five complete pictures each second. If a
shutter of 1/30th is used a complete scan of the picture is not
achieved and the photograph will show the screen with some
areas brighter than others. This effect can be quite marked on
focal plane shutter cameras if a fast shutter speed is used.
The Dutch test pattern clearly illustrates the shading that
can occur with the use of a focal plane shutter. The shading
pattern may disappear at shutter speeds greater than ¼ second.
Experience has shown that the cheaper iris type shutter camera
is ideal for TV-DX photography since it can work down to
1/15th second exposures with perfect results. Care must be
taken to ensure accurate focusing and positioning of the
camera when taking off-the-screen photographs. The normal
minimum focus distance for the cheaper camera is 1 metre
(3 feet) and with the use of a wide aperture the depth of focus
will be relatively small calling for extreme accuracy in gaining
optimum focus of the eventual print.

The basic exposure required for successful television screen
shots is 1/15th at f.8. using a medium speed film such as Ilford

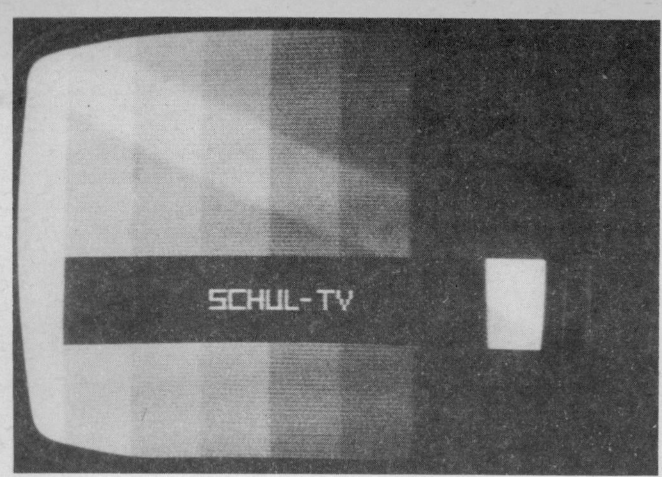

*Shading effect caused by use of a focal plane shutter camera
and too fast a shutter speed*

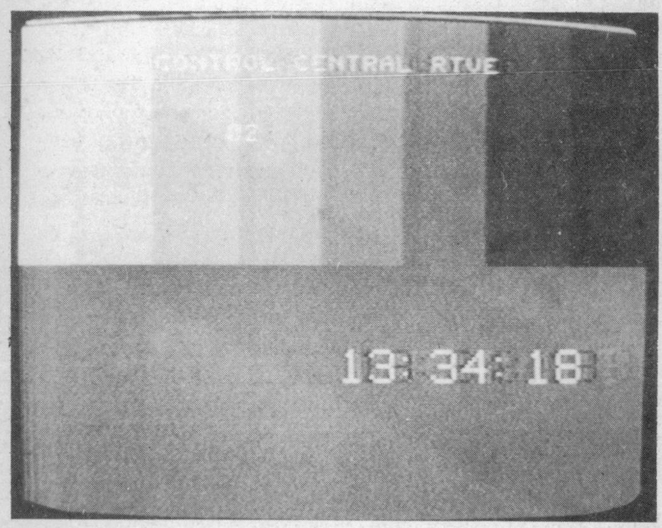

*Another example of shading resulting from the use of a
focal plane shutter camera*

FP4 (125 ASA), or 1/25th at f.5.6. A fast film can of course
be used such as HP4 (400 ASA) but the aperture would need
to be reduced to f.8. or f.11. Some experimentation may be
necessary to determine the optimum settings and film type for
a particular camera and surrounding room environment.

It is possible to use photography in assisting identification of
weak signals using a method known as Time-Lapse Photography.
The usual shutter speed as noted above is about 1/15th second
and for moving subjects on the screen any longer exposure
would result in blurring. Should however a stationary image —
such as a test card, caption, etc. — be received but very weakly
it is possible to remove much of the 'snow' and leave a recognisable picture. Noise is of a random nature and by exposing
the film for up to one second many complete picture frames
will be added together, the noise will tend to cancel out and
the picture information will become much clearer. Since the
long exposure time will allow a considerable amount of light to
fall on the film the aperture must be reduced by the same
factor or a slower film be used. Ilford Pan F (50 ASA) is the
most suitable film using an aperature of f.22 with a shutter
speed of one second. Alternatively neutral density filters may
be used — a filter with a 'filter factor' of x2 will allow the
above camera setting but on FP4 film (125 ASA). There is a
tendancy for blurring when exposing for periods of one second due to sync. jitter and longer exposures will tend to
exaggerate this effect and of course increase the chances of
sync. slip.

When using a camera at such short focal lengths care should be
taken to obtain the correct framing. Often the viewfinder is
located above the lens assembly and appropriate adjustment
should be made to prevent cutting off the top of the frame.
A close-up lens can be used but this calls for extreme care
with focusing distances. If it is possible to occupy the whole
negative area with the television screen, the room in which
the receiver is situated should be darkened, with no direct
light falling onto the screen. However if the screen occupies
part of the negative leaving a dark border, the room should
be illuminated somewhat, allowing surrounding objects to be

Shot 1: exposure of 1/30th second at f.5.6. This shows test card F with only mains details observable, the identification cannot be resolved. Considerable noise is present on the screen

Shot 2: exposure of 2 seconds at f.22 and using a x4|Neutral Density Filter. Ilford FP4 was used for these time lapse shots

distinguished. The reason is that when printing, which is usually carried out with a machine, exposure takes into account the average brightness of the whole negative, tending to over-expose the subject in question. Actual enlargements can of course be made of the subject but these tend to be somewhat more expensive. When photographing from the screen the image should be set for normal contrast settings.

On no account should a flash attachment be used. If at all possible the camera should be mounted on a tripod or other solid object, especially if the shutter is operated manually. Should an unknown test card or caption appear, always resist the temptation of waiting for the signal to improve. It often doesn't! With the use of 35mm film it is wise policy to take several shots of a test card and at a later time the best shot may be selected. Successful screen photography comes mainly from experience and it is suggested that some experimentation is initially carried out with various shutter and aperture sett-ings in order to acquaint oneself with the techniques involved.

STATION IDENTIFICATION

Having equipped oneself for the reception of television signals
over long distances, there then follows the problem of identifying the source of transmissions once the signal has been received.
Fortunately within the European area a large number of
countries have test transmissions throughout the day, commencing programme transmissions during the early evening. The
test cards used during the daytime trade tests usually carry some
form of identification, but unfortunately such cards are often
interspersed with other standard patterns — such as colour bars,
line sawtooths, etc. When such common patterns are used the
only way of identifying the source is to wait the appearance
of the test card or if received during a Sporadic E opening, to
note the general direction, the skip distance and source of
other signals and then determine the transmitter/country being
received by a method of elimination. The situation is further
confused by a transmitter, within a network, originating its
own pattern, whilst other network transmitters originate the
test card. With the multiplicity of channel and frequency
allocations in Europe careful observation of the channel being
received will usually narrow the choice of possible transmitters.
When most transmitters are carrying programme material the
usual means of identification is between programmes when
captions, clocks or announcers appear. At times, within
Europe, a common programme is transmitted throughout
certain countries via Eurovision and/or Intervision. The normal
practice for such exchanges is to display a Eurovision/Intervision
caption indicating the local network, followed with a similar
caption indicating the programme originating network. Often
the Intervision Network captions incorporate well known
local views.

It is possible to identify a programme source from the sound
channel if one is reasonably well versed in languages. Unfortunately many of the East European languages sound alike and as
a general rule such identification is difficult for the person
having little or no knowledge of such languages.

One method of station identification that has enjoyed a degree of success in recent years is by observation of the vertical interval test signal (VITS). If a strong signal from a local transmitter is tuned in and the field hold 'slipped' so that the picture rolls slowly downwards, a series of small lines, dots and other markings will be seen in the field pulse strip. Recently Oracle and Ceefax transmissions in the UK have also been included which will be identified by their continual moving action. The series of small lines and dots are test signals inserted during the field flyback interval in lines 19 and 20. Many European and other broadcasters now insert such test signals and it is possible to recognise individual transmission networks by their particular test signals. The test signals do of course change over a period but with a little practice it will be possible to identify a certain transmitter or country. The test signals are transmitted continually during normal 'test' transmissions and programme times. Several typical VITS patterns are illustrated which will give an idea as to their appearance.

Examples of vertical interval test signals

In the Americas increasing use is being made of offset patterning as a means for identifying an unknown transmitter. To
improve co-channel interference rejection, certain transmitters
are 'off-set"—that is to say: a transmitter will have its radiating
frequency shifted somewhat from the nominal channel frequency by a certain amount, often 10—12 KHz. When two
transmitters are received simultaneously on the same channel
but with one off-set, a patterning will be apparent — caused by
the beating of the two carriers. Having determined the amount
by which the unknown transmitter is offset, reference to
transmitter listings will narrow the field considerably, bearing
in mind such factors as the general direction of reception,
skip distance, etc.

A representative selection of test cards is shown mainly to
illustrate the great variety and types now in use throughout
the world. The whole subject of test cards has been covered
in the publication 'Guide to World-Wide Television Test Cards'
by Keith Hamer and Garry Smith, and published by HS
Publications, 17 Collingham Gardens, Derby, DE3 4FS,
England.

For transmitter identification within the European area the
gether with six bi-monthly supplements. The supplements
advise both opening and closing of transmitters and of changes
in technical characteristics. The 'List of Television Stations' is
published on September 1st of each year and available from
European Broadcasting Union (EBU), Centre Technique, 32
Avenue Albert Lancaster, 1180 Bruxelles, Belgium. Return
postage should be included with any query.

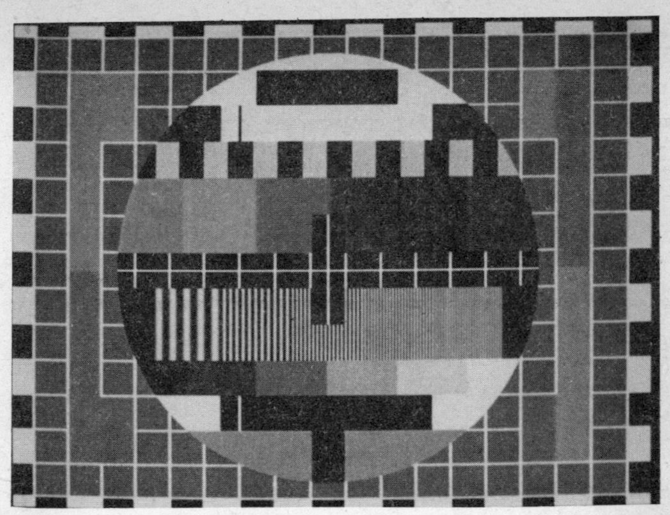

The Philips test pattern type 5544

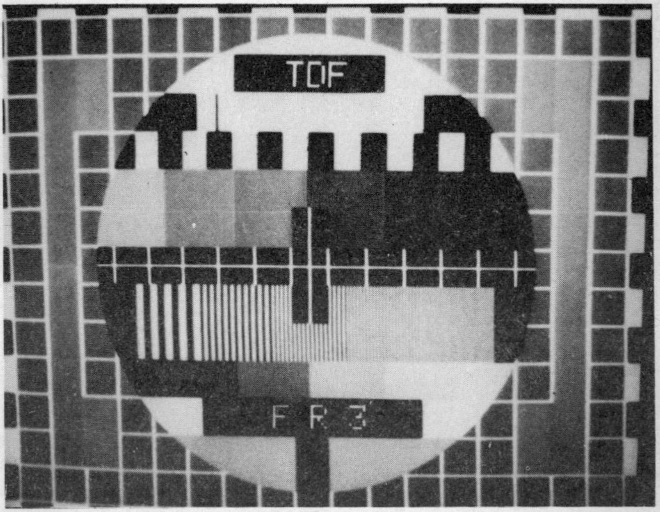

The 5544 in use by the FR3, TDF (France)

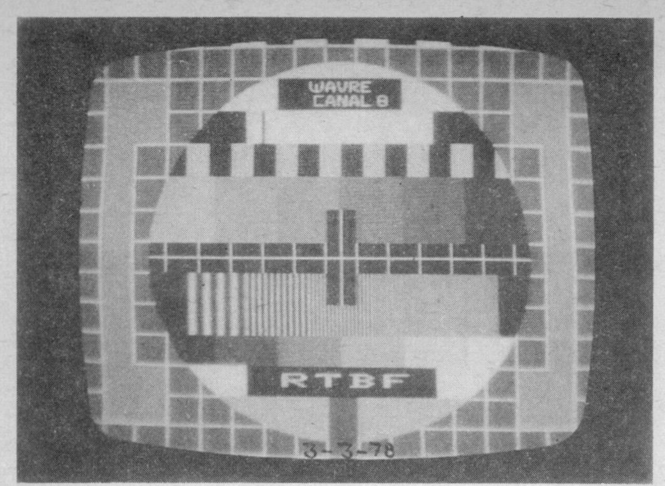

The 5544 in use by RTB (Belgium)

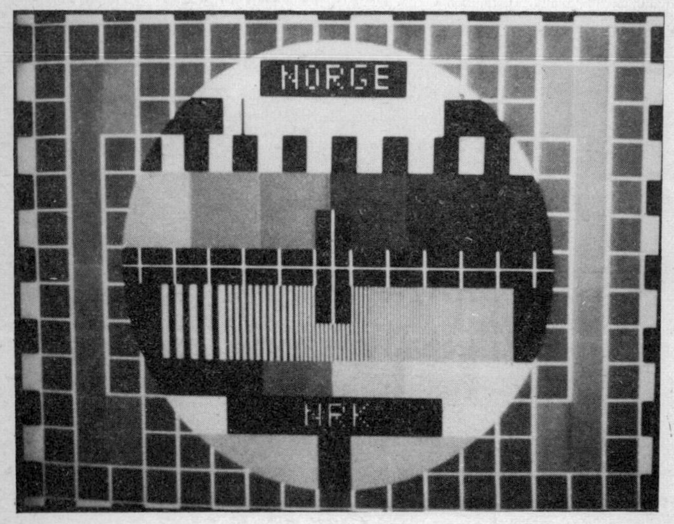

The 5544 in use by NRK (Norway)

118

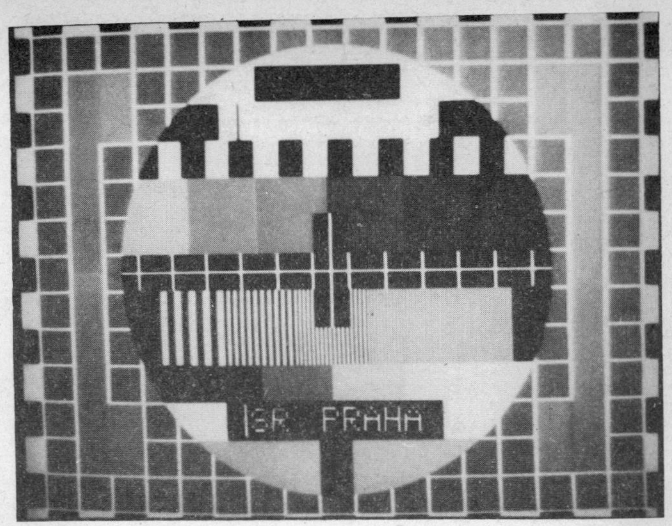

The 5544 in use by CST (Czechoslovakia)

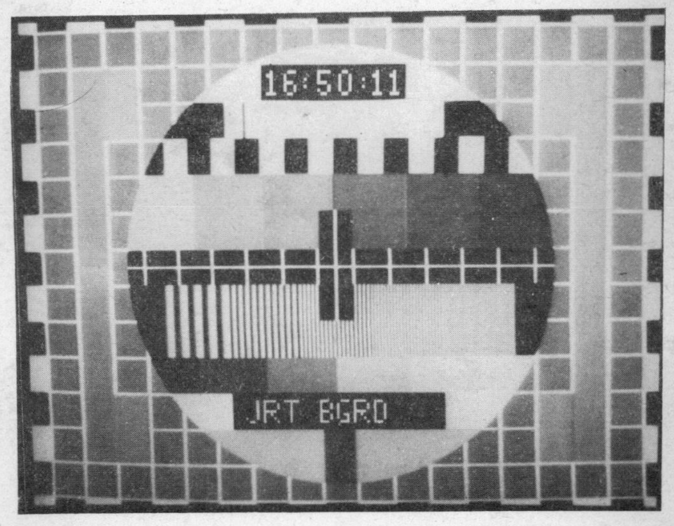

The 5544 in use by JRT (Yugoslavia)

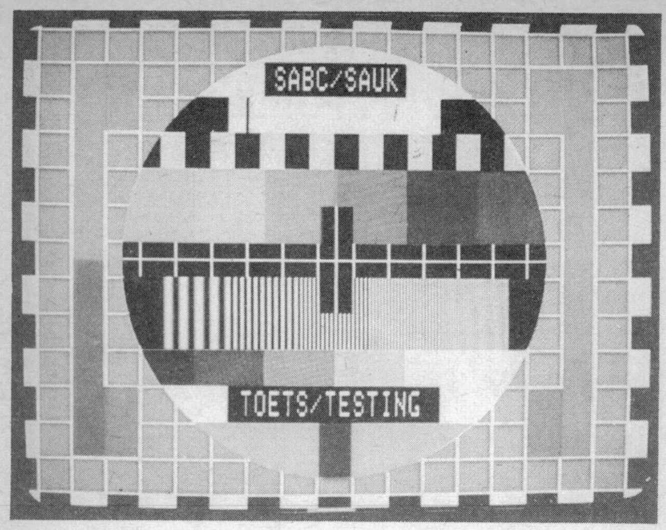

The 5544 in use by the SABC (South Africa)

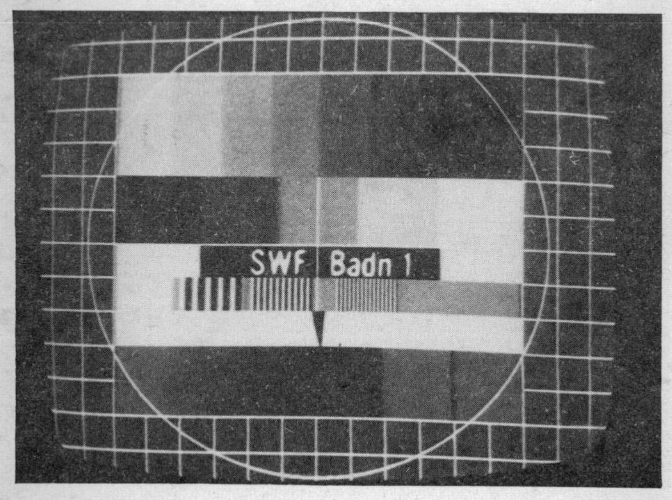

The FUBK test card as used by the SWF (West Germany)

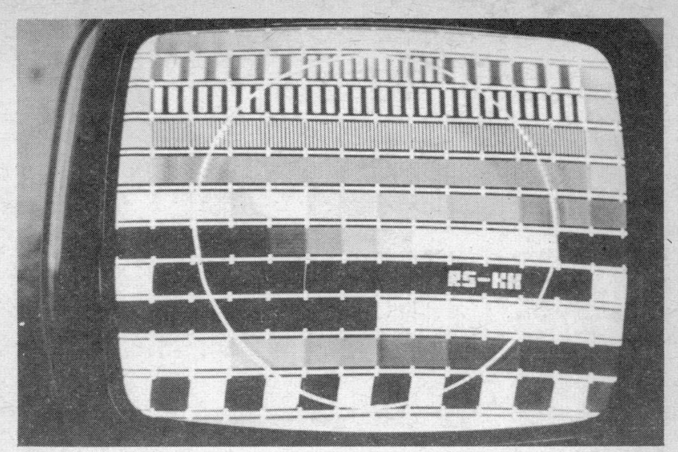

The EZO electronic test pattern in use by the CST
(Czechoslovakia)

The electronic pattern in use by RTVE (Spain)

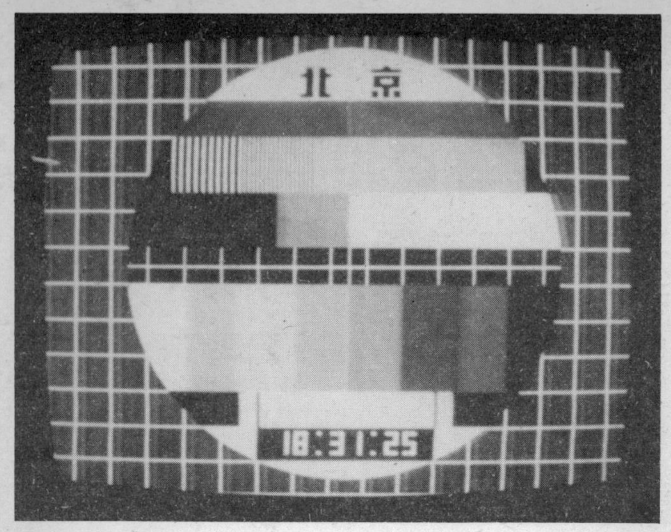

The electronic test card used on the Peking 1st Network (China)

Test card G

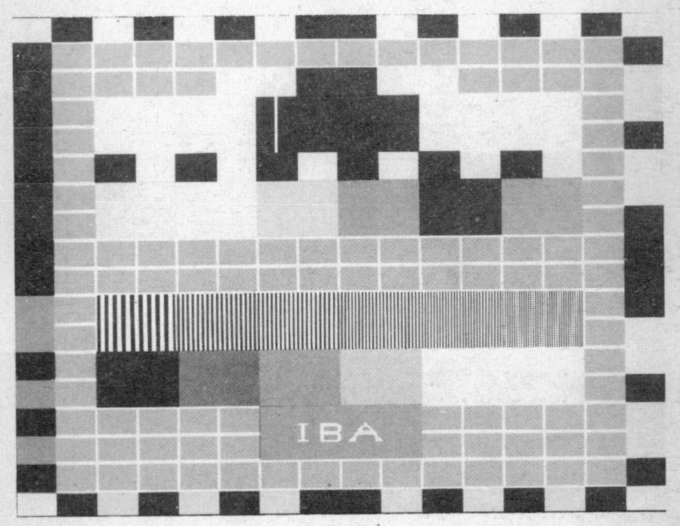

Test card F in use by the IBA (UK)

The IBA electronic test pattern type ETP1 (UK)

Frequency gratings as used by Dubai (U.A.E.)

Station identification slide in use by JTV (Jordan)

INTERFERENCE

With television transmitters in operation in most areas, often
at high signal strengths, some interference must be expected
on frequencies adjacent to occupied channels. The degree to
which such interference occurs will depend upon the strength
of the signal arriving at the receiver from the aerial and the performance of the various sections within the receiver itself. The
question of IF strip selectivity has already been discussed and
where possible the maximum protection against adjacent
channel interference should be incorporated. This will ensure
a constant standard of selectivity irrespective of the channel
in use.

There may be occasions when one particular transmitter causes
problems and one method of reducing such interference to a
low level is with the use of an aerial notch filter. The circuit
of such a filter for use within Band 1 is illustrated, the
response is such to provide an attenuation in the order of
40dB or greater at resonance and allowing signals some 250KHz
to pass with a low order of attentuation. A notch filter of this
type is particularly invaluable when used in the European
area, where, due to the multiplicity of channel allocations,
certain signals may lie extremely close together, an example
being the ch. B3 sound frequency at 53.25 MHz and the ch. IA
video frequency at 53.75MHz. The adjustment of the notch
filter is extremely critical, the tuning must therefore be
adjusted very slowly so as not to pass the notch. When the notch
has been found the variable resistor is adjusted to give a further
increase of attenuation, there being one critical value for this
adjustment. As excellent as this notch filter is, it cannot assist
with co-channel signals; such as ch. E2 video and ch. B2 sound
both at 48.25MHz. In this case more experimentation may be
necessary with a fixed aerial orientated to provide the minimum pick-up of the strong unwanted signal. This method will
give reasonable results with Sporadic E propagation when
signals are strong and with random polarisation.

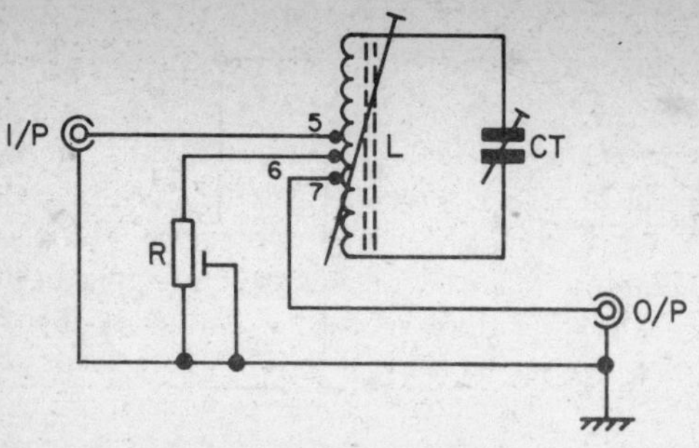

R: 470 ohm miniature preset.

CT: 3-30pf concentric trimmer.

L: ¼" coil former with dust core
11 turns 26g enam. spaced over full
length of former, tapped at turns 5,
6 and 7.

Band 1 (low band) Notch Filter

The design of a double notch filter that features an insertion loss of only 1dB with notch depths typically of 50dB is shown and is intended for applications such as removal of a System A channel's sound and vision carrier frequencies. Experience has shown that attempts to notch out frequencies less than 1.5MHz apart in an attempt to recover a middle frequency does tend to attenuate this latter frequency unduly.

An experimental Notch Filter for Band 3 is illustrated. This was used with great success by an enthusiast living some four miles from a very high powered transmitter in preventing severe cross-modulation and other spurious effects resulting from the strong overload of aerial amplifiers operating on adjacent channels. This design has been used with some success in Band 1. but with the coil being modified somewhat. A Toroid core type T-50 10M1X is substituted and with ten turns of 20g enamel wire tapped at five turns. Such Toroids have a higher Q than the standard core tuned coils due to the reduced number of turns needed.

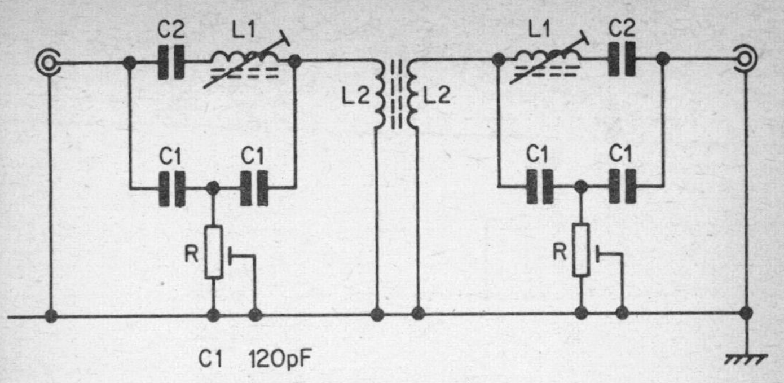

C1   120pF
C2   22pF
R   470Ω min. preset
L1   8 turns 26g ¼" (6mm) former
L2   1/1 coupling transformer
     3 turns each 36g wound through ferrite bead
Notch depth 50dB, through loss 1dB

Band 1 double Notch Filter

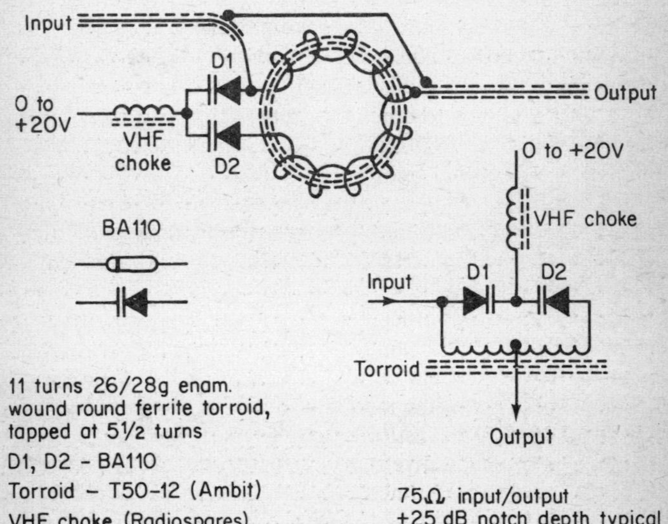

11 turns 26/28g enam.
wound round ferrite torroid,
tapped at 5½ turns
D1, D2 – BA110
Torroid – T50–12 (Ambit)
VHF choke (Radiospares)

75Ω input/output
+25 dB notch depth typical

Varicap tuned Notch Filter for Band 1

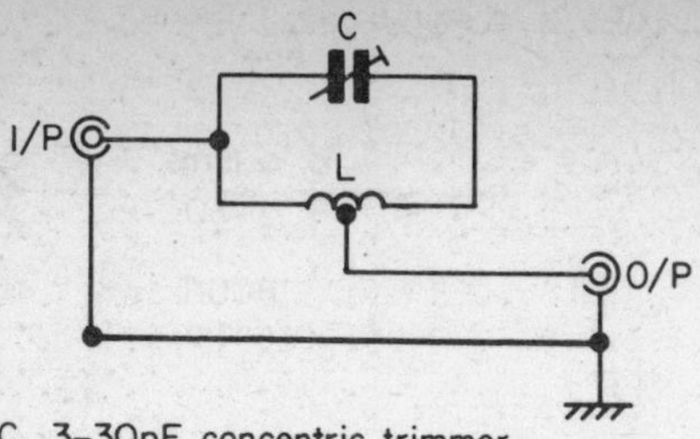

C 3–30pF concentric trimmer
L 3 turns 20g ¼" diam. centre tapped

Experimental Band 3 Notch Filter

In areas adjacent to a high powered UHF transmitter, the latter's signals can often cross modulate into other bands. This can easily be removed by a simple filter at the amplifier input. Such a filter can comprise a series inductance (say 2 turns, ¼ inch diameter) and capacitance (2-10pf miniature low loss trimmer), the circuit being tuned to the mid-channel of the offending group. In severe cases an additional parallel tuned rejector circuit may be necessary in the aerial input (using similar components as before) in addition to the series tuned acceptor circuit. The former circuit is connected between the aerial input and chassis, the second between the aerial input and amplifier first stage.

Reversing the polarisation of the receiving aerial to that of the strong interfering signal can also give a considerable reduction in interference as discussed in the aerial section. Co-channel interference can also be reduced somewhat with the use of highly directional aerials and stacked arrays can be modified to remove unwanted signals from the rear. The phasing harness of such a stacked array has an extra ¼ wave of feeder introduced into one section, and the appropriate aerial is advanced ¼ wave in the forward direction. Signals arriving from the forward direction still combine at the junction of the phasing harness in phase but signals arriving over the back of the aerial are 180⁰ out of phase at the junction (due to one signal having to travel an extra ½ wavelength) resulting in a considerable reduction of the interfering signal.

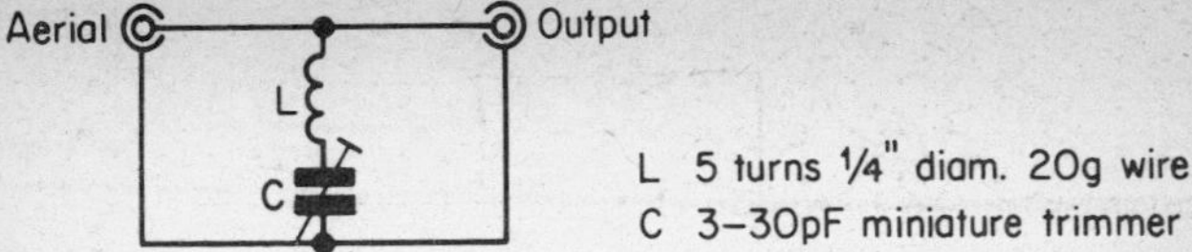

ACCEPTOR CIRCUIT
for removing Band 2 FM radio breakthrough in other bands

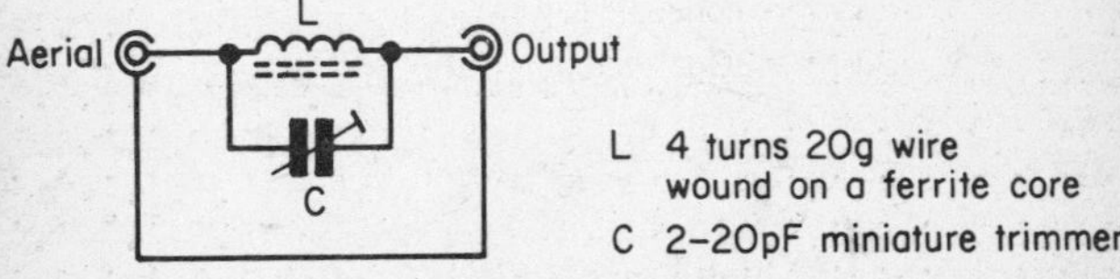

REJECTOR CIRCUIT
for removing Band 3 breakthrough in Band 1
The trimmer is optional and should be used if breakthrough is severe

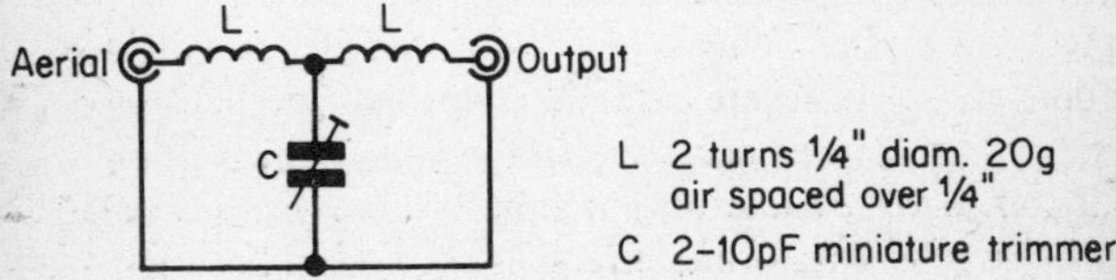

ABSORPTION FILTER
to remove UHF breakthrough in Bands 1/3

NB: The solid inner conductor from low loss coaxial
cable is ideal for making the above coils.

Filters to remove cross-band interference breakthrough

The Radio Services division of the Post Office have a considerable range of filters and if problems are experienced with breakthrough from a known source they can normally be called upon for advice.

It is advisable when a filter is used to remove an offending source of interference that the filter be inserted before any amplifier since the amplifier will undoubtedly exaggerate the problem.

NOTES FOR NON-UNITED KINGDOM READERS

Earlier editions of this volume enjoyed wide distribution and although the general techniques are applicable to most parts of the World it became obvious that certain points needed expansion to foreign readers. As an example the measurements for aerials in this volume are in Imperial (i.e. feet and inches) since despite metrication within the UK, Imperial measurements are still generally used and understood by the majority in preference to the new metric standard.

Metric however is in widespread use in Europe and the following conversions should enable such readers to establish the appropriate dimensions:

> 1 inch = 2.54 cm, 1 foot = 30.48cm., 1cm = 0.394 inches, 1 metre = 39.4 inches., 1 km = 0.621 miles.

Coaxial cable is gaining in popularity due to the ease of fitting and routing. Ribbon feeder however enjoys widespread use, noteably the Americas and Australasia. Most aerial manufacturers have available matching transformers for both VHF and UHF enabling a 75 ohm source (eg an aerial) to convert to 300 ohm ribbon. VHF transformers are inevitably ferrite cored devices but those for UHF may either be of ferrite or a miniature printed circuit board. The transformers are wideband in operation and have only a nominal insertion loss (typically less than .5dB) provided a good quality device is employed. It is possible incidentally to obtain ferrite cored aerial combiners in certain overseas markets which allows 2 aerials to be combined/stacked without the bandwidth restriction associated with conventional cable harness. Again good quality units must be used in the interests of good matching and a low insertion loss. Certain tropical areas are prone to high level static conditions and regular thunder/lightning storms. It is advisable to avoid the use of a masthead amplifier in these areas since a close strike or gradual build-up of static within an aerial system can well mean the demise of a transistor in such an amplifier. It is a wise precaution to include static dis-

charge paths with an aerial system/feeder such as VHF chokes
to earth or high value discharge resistors (eg 5k ohms non
inductive). Certain masthead amplifiers will have protection
diodes fitted across the input circuit to reduce the possibility
of damage due to static build up.

Voltage supplies in certain parts of the World can fluctuate
considerably, if it is necessary to adjust mains tappings
regularly then some means of monitoring the incoming mains
should be provided (eg voltmeter) to avoid overdriving the
components in the receiver. Transformers are available that
will give a relatively constant output voltage between extremes
of incoming voltage. It must again be stressed that TV receivers
can be lethal and due precautions must be taken if the rear
protective cover is removed.

A further danger must be stressed. In previous sections mention
has been made of Sunspots. The Sun must never be viewed
directly with a telescope or binoculars since serious injury to
the eye will result even if viewed for an instant. Attempts at
observing sunspot activity must only be made by projecting
the Sun onto a white card.

CONCLUSION

In the foregoing pages has been discussed the possibility of receiving television signals over long distances, and the resolving of such pictures hopefully with the minimum of distortion on the TV screen. It is certainly possible, and with the minimum of equipment to receive strong signals over distances far in excess of the normal 'local' signals. A beginner to the hobby would be well advised to initially concentrate on the easier modes of propagation such as Sporadic E. This enables signals of high strength to be received over quite considerable distances and with the very basic of aerial systems — a wideband dipole feeding into a VHF Band 1 receiver sufficing. The results, hopefully spectacular, will then enthuse the viewer to quest for the more difficult forms of propagation that necessitate greater skills and improved hardware — and of course greater dedication to the hobby.

The content of this book has been compiled by an active enthusiast, and with the assistance and suggestions from other TV DX enthusiasts, and in the hope that the accumulated information will be a practical guide for the beginner and a source of reference for the established enthusiast. May you enjoy good reception.

TABLES

dB to Voltage Ratio

dB	Ratio	dB	Ratio	dB	Ratio	dB	Ratio
0	1.00	15.0	5.62	30.0	31.6	45.0	178
0.5	1.06	15.5	5.96	30.5	33.5	45.5	188
1.0	1.12	16.0	6.31	31.0	35.5	46.0	200
1.5	1.19	16.5	6.68	31.5	37.6	46.5	211
2.0	1.26	17.0	7.08	32.0	39.8	47.0	224
2.5	1.33	17.5	7.50	32.5	42.2	47.5	237
3.0	1.41	18.0	7.94	33.0	44.7	48.0	251
3.5	1.50	18.5	8.41	33.5	47.3	48.5	266
4.0	1.59	19.0	8.91	34.0	50.1	49.0	282
4.5	1.68	19.5	9.44	34.5	53.1	49.5	299
5.0	1.78	20.0	10.0	35.0	56.2	50.0	316
5.5	1.88	20.5	10.6	35.5	59.6	50.5	335
6.0	2.00	21.0	11.2	36.0	63.1	51.0	355
6.5	2.11	21.5	11.9	36.5	66.8	51.5	376
7.0	2.24	22.0	12.6	37.0	70.8	52.0	398
7.5	2.37	22.5	13.3	37.5	75.0	52.5	422
8.0	2.51	23.0	14.1	38.0	79.4	53.0	447
8.5	2.66	23.5	15.0	38.5	84.1	53.5	473
9.0	2.82	24.0	15.9	39.0	89.1	54.0	501
9.5	2.99	24.5	16.8	39.5	94.4	54.5	531
10.0	3.16	25.0	17.8	40.0	100	55.0	562
10.5	3.35	25.5	18.8	40.5	106	55.5	596
11.0	3.55	26.0	20.0	41.0	112	56.0	631
11.5	3.76	26.5	21.1	41.5	119	56.5	668
12.0	3.98	27.0	22.4	42.0	126	57.0	708
12.5	4.22	27.5	23.7	42.5	133	57.5	750
13.0	4.47	28.0	25.1	43.0	141	58.0	794
13.5	4.73	28.5	26.6	43.5	150	58.5	841
14.0	5.01	29.0	28.2	44.0	159	59.0	891
14.5	5.31	29.5	29.9	44.5	168	59.5	944
						60.0	1000

20 dB = Voltage Ratio of 10 (10^1)
40 dB = Voltage Ratio of 100 (10^2)
60 dB = Voltage Ratio of 1000 (10^3)
80 dB = Voltage Ratio of 10,000 (10^1)
1 dB = Voltage Ratio of 100,000 (10^2)
120 dB = Voltage Ratio of 1,000,000 (10^3)

dBmV Conversions to Voltage Levels

dBmV	Voltage (μV)	dBmV	Voltage (μV)	dBmV	Voltage (mV)	dBmV	Voltage (mV)
−60	1.00	−30	31.6	0	1.00	+30	31.6
−59	1.12	−29	35.5	+1	1.12	+31	35.5
−58	1.26	−28	39.8	+2	1.26	+32	39.8
−57	1.41	−27	44.7	+3	1.41	+33	44.7
−56	1.59	−26	50.1	+4	1.59	+34	50.1
−55	1.78	−25	56.2	+5	1.78	+35	56.2
−54	2.00	−24	63.1	+6	2.00	+36	63.1
−53	2.24	−23	70.8	+7	2.24	+37	70.8
−52	2.51	−22	79.4	+8	2.51	+38	79.4
−51	2.82	−21	89.1	+9	2.82	+39	89.1
−50	3.16	−20	100	+10	3.16	+40	100
−49	3.55	−19	112	+11	3.55	+41	112
−48	3.98	−18	126	+12	3.98	+42	126
−47	4.47	−17	141	+13	4.47	+43	141
−46	5.01	−16	159	+14	5.01	+44	159
−45	5.62	−15	178	+15	5.62	+45	178
−44	6.31	−14	200	+16	6.31	+46	200
−43	7.08	−13	224	+17	7.08	+47	224
−42	7.94	−12	251	+18	7.94	+48	251
−41	8.91	−11	282	+19	8.91	+49	282
−40	10.0	−10	316	+20	10.0	+50	316
−39	11.2	−9	355	+21	11.2	+51	355
−38	12.6	−8	398	+22	12.6	+52	398
−37	14.1	−7	447	+23	14.1	+53	447
−36	15.9	−6	501	+24	15.9	+54	501
−35	17.8	−5	562	+25	17.8	+55	562
−34	20.0	−4	631	+26	20.0	+56	631
−33	22.4	−3	708	+27	22.4	+57	708
−32	25.1	−2	794	+28	25.1	+58	794
−31	28.2	−1	891	+29	28.2	+59	891
						+60	1.00 V

LTS = 10^{-6}

TS = 10^{-3}

Please note overleaf is a list of other titles that are available
in our range of Radio and Electronics Books.

These should be available from all good Booksellers, Radio
Component Dealers and Mail Order Companies.

However, should you experience difficulty in obtaining any
title in your area, then please write directly to the publisher
enclosing payment to cover the cost of the book plus
adequate postage.

If you would like a complete catalogue of our entire range
of Radio and Electronics Books then please send a Stamped
Addressed Envelope to.

BERNARD BABANI (publishing) LTD
THE GRAMPIANS
SHEPHERDS BUSH ROAD
LONDON W6 7NF
ENGLAND